D1825138

Heidegger's Early Philosophy

Continuum Studies in Continental Philosophy
Series Editor: James Fieser, University of Tennessee at Martin, USA

Continuum Studies in Continental Philosophy is a major monograph series from Continuum. The series features first-class scholarly research monographs across the field of Continental philosophy. Each work makes a major contribution to the field of philosophical research.

Heidegger's Early Philosophy
The Phenomenology of
Ecstatic Temporality

James Luchte

continuum

Continuum International Publishing Group

The Tower Building   80 Maiden Lane
11 York Road   Suite 704
London SE1 7NX   New York NY 10038

www.continuumbooks.com

British Library Cataloguing-in-Publication Data
A catalogue record for this book is available from the British Library.

ISBN-13: PB: 978-1-4411-9702-3

Library of Congress Cataloguing-in-Publication Data
Luchte, James.
  Heidegger's early philosophy: the phenomenology of ecstatic
  temporality/James Luchte.
  p. cm.
ISBN 978-1-4411-9702-3
  1. Heidegger, Martin, 1889-1976. 2. Time. 3. Thought and thinking.
  I. Title.

B3279.H49L82 2008
193–dc22

                                                        2008001252

Typeset by Newgen Imaging Systems Pvt Ltd, Chennai, India
Printed and bound in Great Britain by Biddles Ltd, King's Lynn, Norfolk

In remembrance of Reiner Schürmann

Contents

Acknowledgements

I begin with an expression of appreciation for the late Reiner Schürmann, student of Hannah Arendt, who taught and worked in New York City at the Graduate Faculty, New School for Social Research (now New School University) until his tragic death in 1993. Participation in his seminars and lectures had and continues to have a profound impact upon my philosophical questioning and my life, and I regard this writing as a provisional attempt to pick up the 'broken pieces' from his questioning. This has also come to include, in a provisional way, his latest *topos* of questioning which has been incited by his incisive interpretations of the *epochs* of philosophy in *Broken Hegemonies* (published *posthumously* by Indiana University Press, 2003).

I would like to thank the late Charles Sherover for his in-depth feedback on my work and for the generous support he gave to me at a crucial juncture in my writing. I would also like to thank Joanna Hodges for her careful scrutiny of an earlier version of this work. I would like to thank Dr. Raymond Turner, former Dean of the Graduate School of the University of Essex, for his sympathetic support for my work. I would also like to thank Johannes Hoff, Beatrice Han-Pile, Stephen Mulhull, Jim Urpeth, Hubert Dreyfus, Simon Critchley and Richard Bernstein for feedback, comments and suggestions at various phases of my research.

On a personal note, I would like to thank Tamara Al-om and my children Zoe, Soren and Venus for their love, patience and support and who have made all of the work worthwhile.

Introduction

Heidegger's Early Philosophy:
The Phenomenology of Ecstatic Temporality

When questions are raised about principles, the network of exchange that they have opened becomes confused, and the order that they have founded declines. A principle has its rise, its period of reign, and its ruin. Its death usually takes disproportionately more time than its reign.[1]

Prologue: Towards an understanding of Heidegger's 'Sein und Zeit' Project

One of the most significant gestures of the published fragment of *Being and Time* is that a radical phenomenological investigation must have an ontic, factical, 'fundament'. This gesture not only concerns the conditions of emergence for any self-interpretation of the being *for whom being is an issue*, but also intimates the irreducible thrown-ness and embeddedness of any philosophical inquiry. In other words, the pretension that thought can extricate itself from temporality, existential spatiality, etc. – in a word, from *finitude* – is for Heidegger, an illusion 'founded' upon non-original conceptions of existence and being. As Nietzsche expresses in the *Preface* to *Beyond Good and Evil*, such other-worldly hypotheses, in this case of the 'good as such', denies perspective, and thus, life itself. An *honest* phenomenology cannot take refuge in idealist or realist ontologies without forsaking the significance of phenomenology as a desire for the truth of *things themselves*. *Dishonesty* would entail a retreat from the phenomenon into a theory of consciousness and its objects, an escape that suppresses and conceals its own radical temporality.

It is within this horizon that I have approached Heidegger's attempts to articulate a *fundamental* ontology – or radical phenomenology – in the 1920s, and its transmutations to come. For Kisiel, as he explains in his impressive *The Genesis of Heidegger's Being and Time*,[2] it had been nearly impossible to approach any understanding of this period of inquiry due to the fact that the crucial texts concerned were simply not available. In this light, when *Sein und Zeit* appeared, a great astonishment arose in the reading public *at the time*. No one but his students had any scent or taste of the brew Heidegger had been cooking up for so long. Indeed, he had published nothing since his *Habilitation* work on Duns

Scotus in 1915. After nearly twelve years of *written* 'silence', *Being and Time* came as quite a surprise to not only the philosophical community, but also to his students (and perhaps to Heidegger himself). Moreover, as Kisiel traces the genealogy of this work, we can see many variations of its themes and radical changes in its articulation even during the run-up to its composition – a writing which comprised three drafts, 'The Dilthey Draft: The Concept of Time', the 'Ontoeroteric Draft: History of the Concept of Time', and 'The Final Draft: Toward a Kairology of Being' – the *existentialist draft* – written in about a month.

Heidegger's variations, *attempts*, and experiments upon theme and designation underscore the contention that an adequate interpretation of *Being and Time* must be considered alongside his lectures and published works of the period. That which is at stake is the *topos* which gathers together intimations of the projected 'Sein und Zeit' project.[3] The fact that the entire work never saw the light of day – in a formally published manner – underlines not only the *makeshift* character of the published fragment, but also that of philosophical inquiry, *per se*. It is common knowledge that Heidegger was not yet ready to submit 'Sein und Zeit', but that *he had to*, if he was to continue teaching, paying his bills, and providing and caring for his family. Out of this facticity emerges, in a rough and ready way, *Sein und Zeit*, a *torso* inscribing its perhaps illusory hope of *more time* to eventually finish itself. Yet, even though we have been given a makeshift, we are still incited to think and *can* enter into the questions asked by the 'Sein und Zeit' project. And, for such an untimely exploration, Heidegger's lecture courses must come into play, not to supplant the 'masterpiece', but to retrieve, and perhaps, set free a more intimate and spacious *topos* of questioning. An honest philosophy, as it is born out of facticity, will acknowledge the finitude and incompleteness of all works of thought. These myriad 'sources' are indications of the phenomenon that is most at issue, as signs that gesture to the traveller the way to an intimate hermeneutics of existence.

That which has changed for our era of readers of *Being and Time* is the availability of the lectures courses prior, simultaneous, and posterior to *Being and Time*. There is also illuminating documentary material surrounding this period which cannot but help to cast into relief these *topoi* of Heidegger's *world*. In this way, the horizons of astonishment in the face of *Being and Time* are slowly being transformed – but not eliminated – as these do not *ultimately* depend upon the gossip of the Anyone (**Das Man**). At the same time, a differing understanding has also been facilitated by the growth of a quite considerable and increasingly diverse tradition of Heidegger scholarship. Yet, that which is still shrouded in mystery is the *meaning* of the state of incompletion of the published fragment of *Being and Time* – and whether or not his 'failure' to *finish* the work means that he did not 'go all the way to the end'. Indeed, Heidegger's original plan projected a book length to rival Hegel's *Science of Logic* or Sartre's *Being and Nothingness*. 'Sein und Zeit' was however never completed, and as it may never be possible to give a reason why or if it remained *unfinished*, we must instead enter into the place of the question of existence itself. In this way, we will begin to disclose the *passage* of original, ecstatic temporality.

This question provokes us to explore the greater questions of the meaning of the project, of which the published fragment of *Being and Time* was merely a step along the way. There have been intimations by those who have sought to put down their spectacles for a while and become open to the phenomenon of finite existence. There has been serious work on this topic,[4] yet one of the deficiencies of the work on the 1920s phenomenology has been the *archic* position given to *Being and Time*. Indeed, reams have been written on *Being and Time*[5], but usually not in relation to any other contemporary texts or phenomena. Even those who are explicitly concerned with the 1920s seem to have a problem with bringing all of these texts together.[6] Kisiel for instance gives us the impression that all of Heidegger's work of the period is meant merely to lead up to *Being and Time* – in some kind of 'teleology'. In this way, he ends his otherwise great book before any consideration of *Basic Problems of Phenomenology* and *Kant and the Problem of Metaphysics*. Such reticence regarding important lecture courses and a *third* published work *of the period* gives the impression that these are dispensable, that they are merely commentaries on the main work or serve only to unpack what is already there in *Being and Time* (or indeed they need not be read to understand *being* and *time*).[7]

In fact, many of the lecture courses both before and after *Being and Time* not only go beyond the 'content' of *Being and Time* (with respect to the larger 'Sein und Zeit' project), but also seek to criticize or revise various claims or constructions in *Being and Time* itself. While I am not in any way seeking to diminish the importance of *Being and Time*, it would be a vast hermeneutical error to disregard the many contemporary unpublished and published works as mere supplements, when in fact, these seek to 'go all the way to the end' of this project – commitments which were expressed in his lecture courses, before and after *Being and Time*. The centrality that has been given to the latter text creates a circumstance in which *Being and Time* is re-mystified, de-contextualized, or, as Heidegger says in his early lectures, un-worlded. In other words, that which Kisiel has applauded as our ability to *finally* understand this work is tragically eliminated in a repetition of an artificial astonishment in the face of a *magnum opus*. Indeed, Heidegger's great work was never written, but remains in a state of *incompletion*, in fragments. It could be noted that in light of the irreducible facticity of *honest* philosophical inquiry that this, as Bataille writes in *Theory of Religion*, is quite appropriate – philosophy is a building site, always incomplete.

The facticity of incompletion is indicated by Heidegger as Care, as that which is incomplete in itself and as that which attains a *makeshift* unity in the resoluteness of a *being-towards-death*. It is in the singularization of the self, as Heidegger suggests in his lectures on Leibniz, that a world is projected as the *a priori* horizon of original ecstatic temporality. It is on this basis that we can then turn (*Kehre*) towards a metontological *topos* of an-archic *praxis* in which Dasein is open to the truth of being. 'Sein und Zeit' takes its *point of departure* from being-in-the-world, but finds its *beginning* in a radical temporal event. In this light, our point of departure will be the various texts which lie ready to hand, but as we seek an understanding of radical phenomenology, our beginning will be to enter into the question itself.

The morphology of radical phenomenology

Heidegger's 'Sein und Zeit' project, or radical phenomenology, exhibits three originally linked aspects, as indicated in the *Basic Problems of Phenomenology*: reduction, destruction and construction.[8]

Reduction concerns an uncovering, finding the phenomenon of original temporality of Dasein as the prerequisite event for a self-interpretation of existence in its 'public' and singular temporal senses or meanings. Since finite knowing is rooted in ontic being, self-interpretation will not have the meaning of a Kantian self-examination of an immovable Subject, but as a radical phenomenology of the self incited by a 'moment of vision' (**Augenblick**) in which the self comes face to face with its own temporal be-ing without evasion. Amid the event of anticipatory resoluteness, there is a radical singularization of the self, which discloses a sense of *my* own be-ing, and of *my* possibility (which is its most pressing issue). In this light, a fundamental ontology is not an ontology in the traditional sense, but a de-substantialization of a 'what' and 'how' essence into this 'that' of radical singularity, a disclosure of existent be-ing, not via *real* or *idealist* predications, but amid an 'event' of self-disclosure and expression.[9] A *moment of vision*, in this way, is a 'primal' sense of 'theory', *theorein*, an open-ness amid a lived temporality, neither an ancient beholding of *nous*, nor a modernist theoretical objectification.

In that one cannot simply decide to undergo or not to undergo such a questioning, and as there seems to be no 'natural' incentive to exit this world of absorbed familiarity, of average everydayness, we must take heed of the word of Heidegger that it is surprise, breach, event, death, catastrophe, war or a work of art that discloses that which is 'there' – if only provisionally. It is an 'event' of being thrown into 'nothing'.[10] Such a 'moment of vision', as with anxiety and *that* call of conscience, breaks in as strange, conspicuous, amid an average circumspection of everydayness. It appears as a disturbance, a disruption of familiar expectancy, and incites a questioning which tears us out of our absorption in the familiar, and beyond the merely unfamiliar, into the uncanny.

A *destruktion*, as with Heidegger's 'task of destroying the history of ontology' is an operation of re-worlding, of excavating of the temporal origin of the categories, a procedure akin to Nietzsche's genealogical destruction of worn-out metaphors, which live as concepts by hiding their 'all-too-humble origins'. *Destruktion* seeks to dismantle normalizing conceptualities which suppress and overpower the phenomenon of existence and its self-expression. At the same time, a project of re-worlding is a radical cultivating of an originary dimension of self-questioning, which is an exploration of one's existence, set free from the regimentation and imposture of a discipline of a formalist and logical parameters.

Construction concerns the self-expression of existence, a forming of concepts, *logos*, amidst a pre-theoretical and pre-practical *topos*. *Being and Time* expresses this fragility of finite knowing, as it shows a sense of the being of one's self amid being-towards-death, and a chance to 'seize hold' of this sense of one's being in resolute anticipation. The event is a disclosure which breaks in as unexpected disturbance, and is the occasion for questioning the unexamined interpretations of one's being. But, in light of his lectures, his expressive hermeneutics of existence must be seen as an array of provisional and revisable *topoi* of one's own being, intimating a return to the raw temporality of a self-interpretation of being-in-the-world which uncovers one's lived temporality – again and again. Indeed, as I will show, it is always from such an event that 'world' emerges for the first time. Heidegger evokes a 'crisis', in which we are thrown amid the falling of ruination, suppression and erasure. A return to the phenomenon must be a counter-ruination, a cultivation of a fragile *topos* of meaning, self-expression, or as Dilthey sought, an *independent ground* for the cultivation of philosophy in all of its diversity. Life expresses itself, and reflects upon this articulation in its quest for an understanding within the horizons of temporal existence.

Heidegger, for his part, seeks to place temporality (transcendental imagination) at the heart of a phenomenology of factical, lived existence, be-ing (Sherover) – indeed, into the heart of philosophy itself. Such a 'placing' is intimated in his indication of the meaning of being[11] in its own self-projection upon a horizon of ecstatic, original temporality (**ekstatisch**, **ursprüngliche Zeitlichkeit** or **Temporalität**). The radical character of Heidegger's phenomenology is revealed in the first sentence of this study where it was stated that any ontological-existential thinking – or radical phenomenology – as an understanding of Being, will have an ontic fundament, that of existence (**Dasein**). Heidegger seeks to overcome an 'ousiology' that conjures 'in our minds' an image of a being, of a thing or substance, which produces the world and its attributes from out of itself. Such an 'ousiology' suppresses the temporality of existence, displacing the intimate playspace of the self and other selves – which is already a condition in which we fathom the existence of the other 'through a glass darkly' (Paul). Although we are wed to falsity, illusion, concealment, Heidegger indicates an 'openness' amid existence – a *topos*, this place of dis-closure where being *lights up* amid the horizons of our own insurmountable finitude.

In Part 1, 'The Phenomenon of Ecstatic Temporality', I will lay out a provisional sketch for that which is referred to as a 'reduction' in *Basic Problems*. This will be a provisional indication of the phenomenon of ecstatic temporality as the *topos* for a hermeneutics of existence. I will explore the *topos* as the be-ing of Dasein and of the concomitant peculiarity of indication and expression that is necessary in light of this radical singularity. This peculiarity concerns not only the 'conceptuality' of the existentials in *Being and Time*, but also the structure and operation of the inquiry into ecstatic temporality. In other words, since we are concerned primarily with the be-ing of Dasein, as a place for the disclosure of phenomena, and not with any of the derived comportments, we cannot either consider Dasein in the sense of an 'object', or ultimately as a 'technique' or 'operation'. Heidegger will seek to indicate a *topos* for an intimate self-interpretation of Dasein.

In Chapter 1, 'Indications of Ecstatic Temporality', after a preliminary conversation with Schalow with respect to the non-propositional sense of 'truth' for Heidegger, I will begin a sketch of the hermeneutic situation of ecstatic temporality, which exists and can be distinguished from a 'generic', 'common' or 'linear' time. The latter is exemplified by Husserl, in his lecture course, *Phenomenology of Internal Time Consciousness*. This sketch will take place in tandem with a reading of his lecture course, *Basic Problems of Phenomenology*, with respect to the comportment between original temporality and 'common' time.

In Chapter 2, 'An Indigenous Conceptuality of Dasein', I will set forth Heidegger's radical destruction of Husserlian phenomenology in light of its rhetoric of 'going to the things themselves'. Amid the *destruktion*, there will emerge an indigenous expression of 'things themselves', but one not burdened with the ontological baggage of traditional conceptualities. In a criticism of Husserl's thematic field of 'pure consciousness', it will be disclosed that the phenomenon to which there will be a dedicated submission, from which will 'read off' indications, is none other than the finite self in its being as a questioner. For Heidegger, finite existence is the *place* where the phenomenon of the questioner emerges and has its being.

In Chapter 3, 'Temporal Expressions of Being-in-the-World', I will lay out a *prospective* 'rough sketch' of the phenomenological conceptuality of temporal existence traced in Heidegger's *History of the Concept of Time*. This expression is *pro*spective as it indicates aspects of existence without however, as of yet, considering the event of expressive origination for this conceptuality. With a juxtaposition of *History* with Kant's post-revision essay, 'What is Orientation in Thinking?', we will lay out a first approximation of an understanding-of-being, and of an indigenous expression, or indigenous conceptuality, appropriate to a hermeneutics of existence.

In Chapter 4, 'Ecstatic Temporality and the Meaning of Being', in a retrocursive answer to the prospective sketch of Chapter 3, I will trace the understanding-of-being and its 'conceptuality' in light of the treatment of ecstatic temporality in *Basic Problems*. In this text, the *existentiale*, or in the words of *History*, the characters of being-in-the-world, are explicitly disclosed as projections of ecstatic or original[12] temporality, a possibility only hinted at in *Being and Time*.

In Chapter 5, 'Kant's Thesis about Being and Existence', I will set out an interpretation of Heidegger's longstanding meditation upon 'Kant's thesis about Being'. Through a rehearsal of the negative and positive versions of the thesis, I will consider the question of concept formation and the limitations of 'real' and 'ideal' predications as expressions of existence. The implication that arises from Kant's thesis is that the be-ing of Dasein is not susceptible to real (or actual) predication, and that the ontological difference necessitates differing modalities of expression, a *plurivocity* in which this middle-world of *being here-there* can find *honest* expression.

In Part 2, 'The *Destruktion* of Ecstatic Temporality', I will explore the destructive phase of Heidegger's project in the Frieburg work, *Kant and the Problem of Metaphysics*, supplemented by specific references to the lecture course *Phenomenological Interpretation of Kant's Critique of Pure Reason* (1928). I will begin a comprehensive portrayal of Heidegger's interpretation of Kant from within the perspective of the broader question of the 1920s phenomenology of original temporality.

In Chapter 6, 'The Retrieval of Ecstatic Temporality', I will investigate the hermeneutical significance of Heidegger's detour through Kant with respect to the interpretative method of radical phenomenology. I will consider Kant's defenders Cassirer and Henrich with respect to Heidegger's exposure of the ambiguous status of the transcendental imagination betwixt the two editions of the *Critique of Pure Reason*. It is this ambiguous status which casts into relief the *susceptibility* of the Kantian text to Heidegger's interpretation, and his desire to articulate an indigenous expression.

In Chapter 7, 'The Excavation of Ecstatic Temporality', I will trace in detail Heidegger's deconstruction of 'Kant' in *Kant and the Problem of Metaphysics* (1929) through an excavation of the *common root* of the 'stems' of knowledge, intuition and concept, each of which is traced to a originary ground in pure imagination, or, again as a rooting of thinking in that which Sherover indicates as a 'power of imaginative integration'.[13]

In Chapter 8, 'The Articulation of Finite Knowing', continuing my reading of the Kantbook, I will set forth Heidegger's 're-writing' of transcendental philosophy, a phantasy he mused upon in *Phenomenological Interpretation*, in which the transcendental power of imagination is disclosed as the 'formative center of ontological knowledge', a root of the theoretical, practical and aesthetical dimensions. Indeed, at the climax of this lecture, in the last few utterances, Heidegger dramatically reveals the transcendental power of imagination *to be* reason itself – a pure, *sensuous* reason.

In Chapter 9, 'Transcendental Imagination and Ecstatic Temporality', I will consider the many parallels between transcendental imagination and ecstatic-horizonal temporality. I will lay out the resemblances between Kant's three syntheses of pure imagination and the three ecstases of original temporality, casting into relief striking parallels and a translatability of imagination and ecstatic temporality with respect to the analogy between temporality as self-affection (**KPM**) and ecstatic self-projection (**BT**). In this light, I will propose, following Schalow, the resuscitation of a non-idealist interpretation of imagination in the context of a hermeneutics of existence.

In Part 3, 'The *Topos* of Ecstatic Temporality', I will sketch out the 'there' (**Da**) of Dasein through an indigenous expression of this intimate phenomenon of temporal existence. It is here that we will explore the constructivist aspect of Heidegger's radical phenomenology in reference to its topic origin, its power of fulfilment and the specific indications that it issued. I will lay out an interpretation of the existential, indigenous expression in the published fragment of *Being and Time*, notably these existentials of Care, Anxiety, being-toward-death, Guilt and Resoluteness. I will begin however with Heidegger's engagement with Leibniz, who is not mentioned in the original plan for *Sein und Zeit*, in his most explicit deconstruction of *logos*, of statement and its claims to be the 'locus' of truth, in his lecture course, *The Metaphysical Foundations of Logic*.

In Chapter 10, 'The Metaphysical Foundations of Logic', I will lay out Heidegger's demand that a *phenomenological* logic be grounded in the 'matters themselves', which he articulated in his lectures on Leibnizian monadology. In a discussion only hinted at in *Being and Time*, Heidegger engages in a *destruktion* of the doctrines of *judgement* and *substance*, in which he uncovers *logos* and being as distinct comportments of existence, where the former has no priority.

In Chapter 11, 'The "Unity" of Ecstatic Temporality', continuing my examination of *Metaphysical Foundations*, I will begin to disclose the existential and temporal horizons which underlay Leibniz's theory of propositional truth through an analysis of the prerequisite grounding of transcendence in 'world'. Moreover, I will sketch out, in contrast to Henrich's 'Unity of Reason', a 'unity' of ecstatic temporality, as a *topos* of expression, which abides the 'recollection' of its authentic existence (self-remembrance of the self as temporality). 'Unity' will occur for the questioning self amid the remembrance and expression of its own free and finite existence.

In Chapter 12, 'The Riddle of Fallenness, the Building Site of Care and Temporality', I will begin to lay out Heidegger's *topos* of *constructive* indications of the phenomenon of existence. I will begin with the prospective existential of Care, a place of *thrown projection*, fallenness, which is disclosed in the disposition of anxiety (**Angst**), as indicated in Division One of *Being and Time*, but only fully analysed in Division Two. It is, as with Chapter 3, *prospective* in that Care, as being-in-the-world will be disclosed as an indication of existence that is limited in itself and can only acquire its meaning beyond itself in the ecstatic projections of temporality, the subject matter of the next chapter.

In Chapter 13, 'Temporality as the Ontological Meaning of Care', I will examine the 'unity' and temporal sense of this be-ing of existence (**Dasein**) which is founded in a temporal event, in the ecstatic transcendence of the self. I will explore the disclosive characters of conscience, guilt and resoluteness in Division Two of *Being and Time* as aspects of an event of singularization with the character of a 'moment of vision' (**Augenblick**). I will close with the question of the temporality of anticipatory resoluteness and of the implications of such a radical temporality for thought and expression. Ultimately, one listens, one is silent – so as to allow for the self-expression of phenomenon that we ourselves are.

In closing, in 'The Circle of Finitude', I will reiterate the task of radical phenomenology as the disclosure of the radical temporal event which 'grounds' lived existence. Yet, this event, as Krell, after Heidegger, has pointed out, turns into *metontology*[14] which situates one's own self interpretation within the intimate horizons of those of other Daseins.[15] In this way, the *topos* for a hermeneutics of existence could be broadened to include, as Heidegger did in his later writings, cultural expressions such as poetry and art, regarded as indications of the phenomenon of existence (**Dasein**). Such a concern with myriad interpretations of Dasein is already indicated in *Being and Time* in the 'Myth of Cura'. One may wish to pass over this myth as a literary curiosity. Yet, this would be to miss the radical significance and potentialities of a radical phenomenology and turn from fundamental ontology to *metontology*.

Part 1

The Phenomenon of Ecstatic Temporality

What is decisive is not to get out of the circle but to come into it in the right way. This circle of understanding is not an orbit in which any random kind of knowledge may move; it is the expression of the existential fore-structure *of Dasein itself. It is not to be reduced to the level of a vicious circle, or even of a circle which is merely tolerated. In the circle is hidden a positive possibility of the most primordial kind of knowing.*[16]

The task of a phenomenology of *ecstatic*, or *original*, temporality begins with a disclosure of an original phenomenon – the *temporality of dasein* amid its thrown projection of lived existence. Original temporality, is for Heidegger, 'the time of dasein itself'.[17] Radical phenomenology is a self-interpretation, a hermeneutics of existence, an expression of the contours of one's own temporal world. Such a radical hermeneutics indicates the task of 'first philosophy', of a *radically* temporal thought enacted *before* the constructed regions of the objectification of beings. It is from this *before* that Heidegger seeks to 'wrest free' the primitive 'ground' that opens amidst the projections of the horizonal ecstases of one's *eigentlich* – singular – temporality.

The 'ground' is a clearing (**Lichtung**), 'there', 'world', *topos*,[18] a situation of truth and untruth, irretrievably wed together as a place of disclosure and concealment. Dasein is 'in the truth', in the sense of *a-lethea*,[19] but as thrown, falling, it is also in 'untruth', concealed from itself. What Heidegger called a metontology as early as 1928, and which is performed by him after the reputed turning (**Kehre**), is built upon these early excursions into 'first philosophy' – upon his tentative, 'pre-philosophical' clarification of the limits of existence, of being-in-the-world. Heidegger's 'dadaism',[20] as Safranski describes, is an attempt to open a playspace (**Spielraum**) for disclosure and expression, a cultivation of a myriadity of phenomena that exist and which are.

Such 'dadaist' language indicates an ecstatic lived existence, ek-sistence, to which a logic of real and ideal predication must remain blind. In this way, the 'speak' of logic, in its restricted horizons, in its blindness to and suppression of a circle of one's intimate self-interpretation, covers over this specificity of be-ing, and thus, forsakes a love of truth. Heidegger's 'dadaist' revolt incites a retrieval of the question of the *truth* of 'truth'. He is responding to that which he sees as a theoretical and ethical objectification of human existence in the 'humane' sciences. He seeks to indicate and to clarify the radical differences betwixt various understandings and expressions of being, such as that between the 'theoretical' and the 'practical', and his seminal distinction betwixt both of these latter and an understanding and expression bound up with the being of Dasein. In this way, the peculiarity of Dasein consists in the fact that the truth (*alethea*) of its existence is not only projected via a 'phronesis' of circumspective concern, of Care, but is also 'grounded' in the temporal event of anticipatory resoluteness that occurs amid being-towards-death (**sein zum Tode**). The *topos* of being-in-the-world is not in this way equivalent to the 'world' as 'nature', as a totality of things, or of their experiential or conceptual representation. For Heidegger, the 'there' of Dasein indicates instead a 'nothing', as it stands 'outside' and 'beyond' beings as the *prius,* the *a priori* horizon for their disclosure.

In Part 1, we will explore provisional indications (re-duction) of the being of Dasein, and of the peculiarity of indication and expression that is necessary in light of this radical singularity. This peculiarity concerns not only the formally indicative 'conceptuality' of the existentials in *Being and Time,* but also the methodology of an inquiry into ecstatic temporality. In other words, since we are concerned primarily with the be-ing of Dasein, this place in its peculiar *remotion,* and not with any of the derived comportments, we may consider Dasein neither in the sense of an 'object', nor as a 'technique' or 'operation'. We must not imagine moreover that this being-in-the-world is an object for an inspection, or a thing ready to hand for practical deployment and utilization. On the contrary, Heidegger's 'phenomenological destructuring' of theoretical and practical is an attempt to retrieve the being of Dasein upon the original *topos* of ecstatic temporality.

Chapter 1

Indications of Ecstatic Temporality

To be sure, as long as we orient ourselves toward the common concept of time we are at an impasse, and negatively it is no less than consistent to deny dogmatically that the a priori *has anything to do with time. However, time in the sense commonly understood . . . is indeed only one derivative, even if legitimate, of the original time, on which the Dasein's ontological constitution is based.*[21]

Heidegger and the 'Truth'

In his work, *The Renewal of the Heidegger-Kant Dialogue*, Schalow indicates a 'dissonance' in Heidegger's 1920s radical phenomenology, that of a 'rift between judgement and truth'.[22] He contends that this dissonance finds its source in Heidegger's own demarcation of a 'location for truth other than in a proposition'.[23] Schalow describes Heidegger's *destruktion* of Husserl's *Sixth Logical Investigation* as a *shift* from the latter's contemplative relation with an object towards an anticipatory, a pre-reflective and pre-predicative topography of *a-lethea*. This hermeneutic openness amid being-in-the-world is described by Sherover in his *Heidegger, Kant and Time*, as a 'pre-experiential' awareness.[24] The primordial *topos* is neither theoretical nor practical, but is a disclosure of an original phenomenon.

Such an interpretation of truth implies a re-appropriation by Heidegger of the 'Greek' sense of phenomenology as letting the phenomenon 'be seen from itself', which sets in distinction from Husserl's phenomenological and eidetic reductions, and the familiar transcendental logic and 'method of truth' of Kant. Beyond, or beneath questions of 'pure consciousness' and of the possibility of *a priori* synthetic judgements, there is already a prerequisite self-disclosure of beings, an unexpected detour to the '"place" of an object'.[25] Heidegger, in this way, lays out an unfamiliar sense of truth, of a self-disclosure in which the finite self is thrown into an encounter with object-ness (**Gegenstandigkeit**). In this way, Heidegger's interpretation entails a *non-cognitive* understanding, an anticipatory disposition (**Befindlichkeit**), which is projected in the midst of be-ing and its temporal horizons.

In his lecture course, *Metaphysical Foundations of Logic* (1928), this situation of dissonance is attributed to the theory that truth has its seat in a propositional judgement, when for Heidegger, this *logos* must be secondary, dependent upon a prior disclosure of beings, and of Being via beings. From within the containment of 'consciousness', for Heidegger, there can be no access to the 'truth'. Consciousness, judgement and the proposition conceal, and thus, symbolize the false. With this contention, Heidegger has opened up a discussion of an original sense of truth expressed in a discourse of the world, expression in its indicative gesture of disclosing that which is *there*. He states:

> Propositional truth is more primordially rooted, rooted in already being-by-things. The latter occurs 'already', before making statements – since when? Always already! Always, that is, insofar as and as long as Dasein exists. Already being with things belongs to the existence of Dasein, to its kind and mode of being.[26]

In this way, dissonance must not be seen as a fatal discord between being and *logos*, *per se*, but is to be interpreted historically as a temporalization of this genealogy of philosophy itself, disclosed as a *historical* dissonance between diverging orientations, with respect to the meaning of 'truth'.

The dissonance, furthermore, indicates a parallel dispute with respect to the meaning of time and temporality. In a rare diplomatic posture, Schalow acknowledges the similarities between Heidegger and certain aspects of Kant with respect to an anticipatory character of the *a priori*. For Kant, anticipation is clearly in evidence in the 'Copernican Revolution' which is concerned with conditions for the possibility of experience, and not with the leading strings of the objects of nature. It would be in this light that Kant's *a priori* synthesis of pure imagination intimates Heidegger's 'forestructure of understanding' which is an unveiling of the *a priori* as possibility.[27] In each case, with pure synthesis and existential understanding, there is an analogous interplay of the aspects of 'spontaneity' and 'receptivity' in a temporal, anticipatory projection upon a horizon (**Woraufhin**). This interplay lays out the sense of a hermeneutics of existence, in which judgement becomes just another 'variety of interpretation'.[28]

Despite these similarities between Kant and Heidegger, however, Schalow reminds us again that Heidegger's radical temporal analytic is contrary to Kant's own explicit self-understanding of his project, in which he never admitted that imagination is the 'common root' of the 'stems'. Dieter Henrich agrees that it is necessary to comprehend the differences between Heidegger and Kant.[29] One must understand each on his own terms. It is Heidegger's concern for the thrownness of the self *amidst the world*, which suggests a differing sense of truth. For Henrich, Kant considered truth in the sense of essence, or of the possibility of a thing with respect to its defining, conceptual predications. These predications are essential and are not therefore susceptible to temporality, which is considered to be of a merely *aesthetic* character. Essence, in this sense, is akin to the analogy of substance, it is 'what' something is, 'what' it possesses or holds together as an 'identity'. Such a 'unity of subjectivity', grounded in the eternal spontaneity of apperception, forbids any infiltration of temporality into the domain of reason. Indeed, for Henrich, Kant cannot even admit to the existence of an *original* temporality, a phenomenon he must cover over with his delineation of time as a form of intuition and as a formal intuition. His domesticated or common time will never be a danger to the unity of reason.

For Heidegger, who does not incarcerate temporality within a transcendental aesthetic, truth is disclosed as a phenomenon circumscribed by the limit of an ecstatic horizon of original temporality. In other words, the truth of the finite self is the self-temporalization of its world. Self-temporalization **is** original temporality, prior to time as a pure intuition, common time and prior to conceptual determination. In this light, so as to disclose the specific character of original temporality, I will turn in the next section to an interpretation of the meaning of a conflict betwixt original and common times in relation to the previous difference between truth as a-lethea and truth as the predication of a subject in a proposition.

The sense/meaning of ecstatic temporality[30]

Towards the very end of the 1927 lecture course, *Basic Problems of Phenomenology*, Heidegger undertakes an extensive discussion of the meaning of original, ecstatic temporality in its distinction from a common understanding of time. For him, original temporality serves as the condition for existence and its expression, not as a substance for an accident, but as a horizon for the disclosure of existence. It is indicated, expressed as characters of temporal being-there, which are intimately 'read off' of the phenomenon, and are guided by it. As Kisiel describes in his *Genesis*,[31] 'original temporality is sense, ground, and condition for a being-in-the-world, "there"; it is the "root" of Care.' It is the sense of being of Dasein's 'whole', as it is disclosed in a 'glance of the eye' (**Augenblick**). Such a temporality departs from the inconspicuousness of the everyday into a breach, as for instance, with the anxiety of being-towards-death. The elusivity of a play of original temporality, an ecstatic temporality of factical lived existence, disrupts stable meanings, fixed identifications – it is itself a questioning which is 'on its way' to the 'matters themselves'. Original temporality points to the temporalization of this self amid its world, in which the finite self enters into the question of the meaning of its being, its most pressing issue.

Heidegger contends in *Basic Problems* that Aristotle's definition of time, although it outwardly resembles Kantian time in being a sequence of 'nows', must instead be regarded as a time constrained by these 'matters themselves', as a counting which follows the movement of the phenomena. In terms of Heidegger's method of formal indication, his reference to Aristotle gives us an 'access definition' or 'access characterization' of time that orients our initial pre-sense of temporality, and which 'corresponds to the common pre-scientific understanding of time'.[32] This is not the first time Heidegger has turned to Aristotle's notion of time. He had already lectured upon Aristotle's theory of time in 1921–22, lectures that would lay out a backdrop for his 1924 lecture to the Marburg Theological Society 'The Concept of Time'. It must be emphasized that the significance of his investigation, just as with the other philosophers with which Heidegger engaged, does not lie in any attempt to refute Aristotle. Instead, he is seeking to trace the rooting of the common concept of time to a more radical, original sense of Temporality. Indeed, Aristotle's notion of time as a counting of phenomena at the very least implies some relation with things themselves, without however allowing the phenomenon to speak amidst the din of numerical succession. It is only when counting is stopped that the phenomenon can express itself.

In his early lectures on Aristotle, a topic which *Being and Time* specifies as an integral aspect of the 'Sein und Zeit' project, Heidegger describes the common sense of time as the number of motion with respect to the horizon of 'earlier' and 'later'. Number simultaneously intimates contingent and permanent aspects of time. On the one hand, number is counted in succession, there is a chain of present moments, 'nows'. Yet, on the other hand, the now is always that which is there in its 'unity', with the withdrawal of the no longer and the reticence of the not-yet. There is a distinct parallel between these aspects of time and that of being as such. For Aristotle, being was *ousia* (ουσια), substance, that which remains a fixed possession amid incessant coming to be and passing away. It is the latter *parousia* (παρουσια) which is the contingent expression of the attributes of substance. In this way, we can see an obvious connection of time and being in Aristotle.

In these early suggestive lectures, Heidegger reveals his suspicion that there must exist a deeper sense of temporality, a primordial or original temporality, which first allows Aristotle's definition of time to have meaning as such. Most pointedly, Aristotle deploys temporal characteristics to define time, and thus, intimates the original *situs* of temporality, serving as the background for his articulation. Moreover, if there is a more original sense of temporality intimated in Aristotle, there must also be the possibility of a radical re-envisioning of the sense or meaning of Being. Time and being(s), conceived as substances, as entities, are for Heidegger derivative or secondary expressions for the existence of things in the world of Dasein. Yet, it is Aristotle's attentiveness to the *motion* of the substances that intimates Heidegger's search for a meaning of being that would be appropriate with an original sense of Temporality. In this light, Aristotle's 'theory' of time will be subjected to a *destruktion* that seeks a non-ousiological, non-entitive sense of Being, as suggested in *Being and Time* and the other texts of the 'Sein und Zeit' project.

It is in *The Concept of Time* where Heidegger first sketches a *morphology* of original temporality. Yet, he does not simply posit a new 'theory' alongside the prevailing hegemonic theory, that is, Aristotle (or Kant). Instead, he undertakes an excavation of the original phenomenon that lies suppressed by the predominant fantasms.[33] He begins with 'clock time' which is at once our own pre-understanding of time and the character of scientific time. In a different way than he will deal with this conception of time in *Basic Problems*, Heidegger invites his audience to enter into a phenomenological investigation of the clock hanging on the wall and the motion of its hands. We sit outside of this clock, under it, watching it enact a spatial repetition of the same, a round and round depiction of numerical time. As long as the clock is wound up or the battery is charged, it 'tells the time'. Yet, we can turn away from this time, from the clock. It is not essential to our existence. We concern ourselves with it only when we wish to know the common time available to everyone – the time by which we synchronize ourselves with others amidst a regime of command and control – of discipline, to use Foucault's suggestive language.

Yet, we already sense that the time of the clock is somehow alien to us, though necessary to our everyday fulfilment of promises and duties. The homogenized repetition of this circular linearity fails to speak to us – we forget such time amid our own temporal existence – a time away from time. We notice suddenly that this linear time has moved along quicker than we have imagined – we are shocked out of ourselves and our involvements to notice *what time it is*. Yet, this is not the only way we measure time or have a sense of time. Heidegger, slowly moving away from the 'age of mechanical reproduction' (Benjamin), mentions the sun dial and the fluctuations of night and day, etc. Each of these indicates ways of telling time, which is accessible to each and all, but ones which perhaps lack the onerous precision of mechanical or digital clock time. Yet, these ways still lie only upon the surface of a deeper sense of temporality which he seeks to dig up, out from under the homogeneity of generic time. Heidegger begins to suggest a temporality of one's own existence, the temporality disclosed as one begins to fathom the singularity and finitude of one's be-ing. The clock on the wall certainly does not speak in any specific sense to the singularity of one's care, to one's involvement in the world, which is an immersion that often leads one far away from any sense of linear mechanical or even terrestrial/celestial time, as with existential spatiality. One forgets the time amid the temporality of involvement, of Care. Clock time recedes in the wake of one's singular enactment of existence. Moreover, beyond even the indication of Care is the issue of the singular finitude of one's existence. The seeming eternality of the mechanical wheel of time and the fluctuations of night and day suppress the radical finitude of existence. Original temporality is disclosed for the existent being in the shattering disclosure of one's overwhelming and insurmountable possibility. Amid the anxiety of being-towards-death, one is given the sense of one's own temporal specificity, which does not merely go round and round, but will stop. There is a breach, a nothing that hangs over each like a cloud. The singular temporality of one's existence and its anxiety, which as a mood has disclosed the horizons of this existence, evokes a differing sense of time for the one whose being is 'at issue'. Original temporality is clearly distinct from that time of the machine on the wall – or, the fluctuations of night and day. Moreover, with this insight that breaks through with anxiety, there is an intimated sense of conscience, a call of the self back to the singularity and specificity of itself – away from the indistinct homogeneity of the Anyone (**Das Man**) and of its concealment of the truth of existence.

In *Basic Problems*, Heidegger states that the common, or generic, understanding of time, 'by its own phenomenological content . . . points back to an original time, temporality'.[34] He illustrates the trace of an original temporality in the content of common time through a consideration of the seeming enigma of clock time: in the clock is 'embedded' the presupposition of the common understanding of time – time as pure succession of nows, a sequence of 'nows', as an infinite, uni-directional sequencing which elapses. The clock brings us to a reckoning with the specific time of ongoing actions, that of an 'in order to'. Our involvement with the clock, Heidegger states, is rooted in and springs out of 'taking time into account'.[35] This 'taking time into account' is a 'guiding oneself according to it', which suggests a phenomenological or 'original comportment with time', as we saw in the case of Aristotle. Time – and our reckoning with time – which is 'somehow already unveiled', is more original than a clock. Such equipment or gear (**Zeug**) is only a symptom or indication of beings who reckon with time. Clock-time refers to the 'wherefore' and 'whereto' of the 'now' of a being amid a nexus of involvements. A 'now' is thus not a substance, or an object, or extant as a monad, but is an expression that points out, refers to, strangely, the one who reckons.[36] Heidegger states,

> The *Dasein*, which always exists so that it takes time for itself, *expresses itself*. Taking time for itself, it utters itself in such a way that *it is always saying time*. When I say 'now' I do not mean the now as such, but in my now-saying I am transient. I am in *motion in* the understanding of now and, in a strict sense, I am really with that *whereto* the time is and *wherefore* I determine time. However, we say not only 'now,' but also '*then*' and '*before*.' Time is constantly there in such a way that in all our planning and precaution, in all our comportments and all the measures we take, we move in a silent discourse: now, not until, in former times, finally, at the time, before that, and so forth.[37]

The 'then' indicates a being-expectant, 'at the time' refers to the 'bygone', which can also be a forgetting. In the explicit or tacit utterance of 'now', 'I' comport myself to an 'enpresenting,' to a having-there amid my 'present'. This intimate nexus of indications is expression, as a 'self exposition of comportments'.[38] Time expresses itself through these comportments, and this coherence or 'unity' of time is indicated via the 'enpresenting implicit in every expecting and retaining'.[39] This inner coherence, similar to the *sketch* of imaginative synthesis, as we shall see, is a 'coherent connection' of '*time in a more original sense*'.[40]

The time of a clock points to a time of significance, of the world in its factical everydayness: of 'in-order-to' and 'for-the-sake-of', 'for-that-purpose', 'to-that-end'. Such comportments configure a 'world-time'. Being-in-the-world is a temporal clearing; it is not extant, but exists as an enabling of an 'uncoveredness'. What is uncovered with this uncoveredness is a *topos* of temporal indications, which exhibit an inherent datability, of a 'now-when', 'then-when', 'at-the-time-when'. Datability implies an 'event', 'work of art', or thing 'of the world' that discloses that which is *there* as the phenomenological content of a 'totality of significance'. The temporal indications, or determinations of time, as with the schematism of understanding in Kant, are joined to that which 'gives a date to the datable'.[41] Amid this plurivocity of existential datability, there is a stretching out of time that is its *spannedness*, the opening of the place of the mutable, transitionary. Spannedness coalesces 'in our being with one another', and thus this temporality is '*publicness*'. Time is here for every one, but, at the same time, each is her/his own peculiar datability vis-a-vis his/her own existence. Heidegger echoes Nietzsche lament of the Last Man in *Thus Spoke Zarathustra*: 'The now is accessible to everyone and thus belongs to no one.' A *peculiar objectivity* is born. One's time is to be taken or to be lost in the 'oblivious passing of our lives'.[42]

Temporality, in this way, and contrary to an abstract picture of time, expresses itself through our being-in-the-world and the be-ing of the self. In expecting, this being-there of the self is ahead of itself. Comportments of expecting are oriented to an ecstasis of futurity. Heidegger describes, 'I come from this possibility toward that which I myself am.'[43] The self exists as its thrown projection, one comes towards oneself from futurity, one projects oneself upon one's 'potentiality-for-being'. The primary concept of the future is thus a 'coming to oneself from one's most peculiar possibility'.[44] In retaining, on the other hand, being-there is contained in what it has already been, a having-been that belongs to its own future and which *bleeds* into that which we commonly refer to as our lived present. Heidegger insists:

> The Dasein can as little get rid of its [past as] bygoneness as escape its death.[45]

In an existential sense, these determinations of time are prior to average time, and the original 'unity' of these determinations point towards an original, ecstatic temporality. Temporality temporalizes/expresses itself. Heidegger writes:

> In expressing itself, temporality temporalizes the only time that the common understanding of time is aware of.[46]

Amid this temporalization, these expressions of lived temporality become: coming towards oneself, going back to, staying with/dwelling with. The 'toward', 'back to' and 'with' are the indications that lay out for Heidegger the basic constitution of temporality.

Temporality as a play (**Spiel**), a 'unity' of future, past and present, is 'outside itself'. It does not carry us away merely from 'time to time', but, is always outside itself. Heidegger traces this character of temporality to the ancient Greek *ekstatikon* (εκστατικον), shedding some light upon the meaning of an 'ecstatic' or original temporality. Each ecstasis exhibits a 'co-equal originality', or as Henrich suggests, an 'equi-primordiality'. Temporality, as the *topos* of existence, 'within itself . . . is outside itself'. Amid this horizon of the ecstasis, there exceeds a 'peculiar openness'. Ecstatic-horizonal temporality 'opens up this horizon and keeps it open'.[47] Yet, despite his intimation of this peculiar openness, Heidegger articulates the necessity of offering a proof for his indication of an original temporality. He states that such a proof, similar to Kant's Transcendental Deduction, must show how common time is rooted in original temporality, or, from the obverse, how original temporality makes possible the common time which '. . . factical dasein experiences?'[48] Heidegger writes in *Being and Time*:

> The concept of 'facticity' implies that an entity 'within-the-world' has Being-in-the-world in such a way that it can understand itself as bound up in its 'destiny' with the Being of those entities which it encounters within its own world.[49]

Heidegger asks after the possibility of a self-understanding of original, ecstatic temporality – from amid the absorption in the destiny of things, of beings. Such a possibility, if it can be disclosed, will allow for an understanding of the rootedness of common time in original temporality. Yet, this question announces itself amid a facticity in which the ecstatic-factical lived existence is covered up, normalized/erased in the routines of common time, to an addiction to an infinite sequence of 'nows'. The be-ing of the self, if the question is asked, is interpreted in light of a tacit sense of the being of beings, which means that the being of the self is projected in the manner of an 'ontology of things', of technic and principle, and thus, in the light of common time. In this way, intimate temporalities of the self and of existence are suppressed in a self-forgetting, concealment. The self becomes once again captivated by the shadows of the 'cave', in the *polis* of Anyone, measuring shadows which dance upon that wall – sketching patterns and repetitions. Yet, as Heidegger intimates, such a forgetting is not fixed, but has its own temporality of reversal.

The return to original temporality

In *Basic Problems*, as we have suggested, Heidegger traces the 'existential' rooting of characters of generic, or common, time to original temporality. Such a tracing, or genealogy, is akin, as Han-Pile suggests, to the search for conditions of possibility, in that it operates according to *a priori* procedures through which the phenomena is anticipated through a 'pure knowing'. Heidegger's insistence upon the 'common root', however, distinguishes his interpretation of transcendental philosophy from that of Kant. The explicit difference lies in Heidegger's disclosure, excavation, of an original sense of truth. In order to understand this 'original sense of the *a priori*', we must trace Heidegger's derivation of the characters of common temporality, in this case of datability, significance, spannedness and publicness from original temporality. In a peculiar way that will become clear in the later chapters of this study, these characters are 'figures' of generic time, but characters which only become intelligible in light of ecstatic temporality, beyond/before generic time. Heidegger's derivation will seek to disclose the temporalizing of temporality into 'now time'. He will further ask *why* original time is 'proximally and for the most part' covered over or overlooked by the factical self who primarily experiences this 'now time'. How is this possible if the time of the now is 'temporality expressing itself'?[50] Or, perhaps, the better question would be why would there be the suggestion of an original temporality at all, differing from the generic, familiar understanding, unless there were already the occasion for the question? What occasion is that in which we would break out of our absorption in the ordinary, in the cave of things? Captivated in this labyrinth of consumption and production of things, why would anyone ever think differently?

In *Basic Problems*, Heidegger sets forth (in retrospect) a derivation of generic time from original temporality: ecstatic temporality temporalizes 'now time', the hegemonic time of the clock. The now is a 'now' only when it is dated via a being or event – 'now when this or that'. The 'now' that exhibits the character of *datability* is 'ecstatically open'.[51] The 'now', in a way which will become clearer in Chapters 3 and 4, belongs to this ecstatic-horizon of the 'present', as an 'enpresenting of something'. Towards this something the 'now' is ecstatically open. 'Now', in this way, is a self-expression of the 'present', it opens the 'present' as a 'related to-', a *topos* where a being can be encountered. An expressed 'now' dates itself with this being/event that it encounters as it announces/expresses itself in the Open. The 'now' abides in a 'structure of enpresenting', the temporal character of which points towards an ecstatic temporalizing of time that 'already lets beings be encountered as uncovered'.[52] This dating-indication sheds light upon the common understanding of time in that the 'now', instead of being a free-floating point in an infinite and irreversible *ether*, becomes an indicator, index, of existence amid an understanding rooted in temporality.

Each time-determination is also spanned, or exhibits *spannedness*, an 'expression' or 'speaking out' rooted, for Heidegger, in original temporality. Time as *stretchedness* is a temporalizing; it is ecstatic, 'stretched out within its own self'.[53] However, this stretch is 'basic', 'primordial' – it is not an aggregate, or a bridging-together of disparate items, but an original unity which precedes the severance of theory and its alleged *manifold*. It is the self-expression of temporal existence. In this way, each time determination is 'publically accessible' to the understanding of the self and others, the being here-there of the self and its world ('being there with one another').[54] The indication of publicness also traces its roots to an original, ecstatic temporality that is 'open for itself along the directions of its three ecstasies'.[55] Openness, as we have seen, is a temporal world, *topos* of significance, meaningfulness. The self-expression of time that is the self-expression of our own existence, is known to everyone in the 'communal being-in-the-world', in this 'unity' and *openness* of an original temporality (historicity). Expressed time is 'world time', a time which is bound up in the destiny of Dasein, using itself for itself, occupied within its own time of existing. The self amid its being in its world, among others, this belonging together of Dasein, is the 'temporal entity simply as such'.[56] Since Dasein is 'intrinsically temporal' in an 'original, fundamental way', it stands outside of the 'frame' (**Gestell**) of common time, 'out of time', although remaining 'in time'. Existence and its finite dissemination, the lived existence of Dasein, is a thrown projection that is the prerequisite openness for extant things in the 'world'. There is a difference in its being which is intimated/hinted at and can only become clear as we step back towards ecstatic temporality.

The 'covering up' occurs as a falling where Dasein is primarily oriented to an extant being 'in time'. It understands its own being in terms of entities, and has done so historically, as for instance as ego, subject, *res, substantia, subjectum*, etc. Any self-interpretation or understanding, as it lives amid these horizons of the extant, 'receives the determination of extant being'. One understands oneself in terms of the things and horizons which press against it. In the facticity of clock time, temporality is connected to 'things', and is 'concomitantly extant' as a sequence. The 'now' becomes *clothed* as an intra-temporal 'intrinsically free-floating run off of a sequence of nows'. This welter of 'nows' fabricates, via its linear *mathesis*, an illusion of everlastingness, although in truth, it is finite, and soon *breaks down*. Any 'now' becomes the being of time, that sequence with a transitionary character of a 'not yet' and 'no longer'. A 'now' is slippery, always points beyond itself – it promises an 'everlastingness', but this bad infinity of a 'clipped sequence of nows' covers over our true 'eternity'. The endlessness of generic time arises for Dasein since temporality, intrinsically, forgets its own essential finitude. Ecstatic temporality is 'further outside' than any temporal object or position, and is, Heidegger describes, the 'self-enclosed ecstatic horizonal unity of future, past and present . . .'[57] Original temporality lays out a *topos* for understanding, and is thus the condition of possibility for the existence of Dasein and of its understanding-of-being. But, once again, how is such an understanding possible? From *where* does the question emerge? *Where* does the questioner notice itself as a questioner? Once again, why would, could, anyone ever think differently?

The *Destruktion* of generic time

To cast into relief specify Heidegger's tracing of the characters of generic, common time to our *topos* of existence, of original temporality, in *Basic Problems*, I will turn to Husserl's treatment of time in the *Phenomenology of Internal Time Consciousness*. For Heidegger, this text is a significant example of a now-time which owes its origin, along with the 'mystical' conception of infinity, to the transplantation of a mathematically constructed time, a *mathesis universalis* into phenomenology. This genealogy provokes Heidegger to declare that Husserlian phenomenology was dead in the water, a still birth that failed in its inception. The significance of this engagement between the two thinkers is underscored by the fact that Heidegger edited and compiled this book of lectures. In Sections 32 and 33 of *Phenomenology of Internal Time Consciousness*, Husserl writes that he seeks a 'consciousness of a unitary, homogenous, Objective time'. He surmises that we can fathom such an Objective time if we enact a regression into the past via reproduction/phantasy, as a 'successive stringing together of temporal fields'. This regression strings together past intervals, preserved 'objectivities', together which form a linear order up to the prevailing 'now'. Husserl writes:

> Even every arbitrarily phantasied time is subject to the requirement that if one is able to think of it as real time (i.e., as the time of any temporal Object), it must subsist as an interval within the one and unique Objective time.[58]

For Heidegger, generic 'now time' is the time of theoretical levelling – it falsifies, covers up the singularity of factical, lived existence, a temporality which disappears in 'objective time'. Such time is instead, for Heidegger, founded upon an original temporality of simple apprehension. 'Now' or 'generic' time renders everything familiar, 'safe', defined. It covers over this 'uncanniness' of ecstatic temporality, but yet, as a covering up, it is a forgetting somehow 'built-in'. Despite the explicit eruption of a direct apprehension of 'matters themselves' – for instance, with being-towards-death – such a disclosure cannot be immediately accessible 'all the time'. Over time, it becomes forgotten, displaced, actively covered up, or, perhaps it is simply absorbed back into a levelled, falsified temporality of 'now time', that time of the clock. Clock time is not a temporal expression of lived existence – it is a mechanical time of factical effect, it is an automaton which incessantly forgets. For Heidegger, it is blind, ceaselessly distracted from the situation of existence, absorbed in the rhythm and flashing lights of reified things, of a 'one size fits all'. Here there is a universal and particular, but no *singular* existence.

Husserl sets out an objective, linear, temporal order with significant implications for the question of the relationship betwixt temporality and thought.

In the first instance, Husserl's insistence upon a strict identity allows for the emergence of an extra-temporal 'now time'. This time, linear, infinite (theoretical), composed of discrete times within a single unique time, is founded upon, or constituted by, an 'absolute consciousness'. It becomes a linear order of unalterable temporal positioning, grounded upon the identity of projected material domains, as with his example of the simple C tone. For Husserl, as with Aristotle, time is an infinite two-dimensional line of discrete identifiable moments which stream off into a past in 'diminishing returns', a familiar notion, which lets the *new*, unfamiliar, be assimilated into the *old*, into the past. But, as it is posited in the present, it always shines with the seductive intensity of the *new*, of an 'unlimited'. Yet, Husserl maintains strict distinctions: the past is past, the now is the edge, the limit of the past, the future will come out of the past via the present.

Husserl begins with an emphasis upon the 'lived experiences' of time. Yet, by refusing to pursue the question of the *temporality* of 'lived experiences', he positions 'lived experience' within the frame of 'consciousness', itself unaffected by time, but still somehow *grounding* temporal 'objectivities'. The lack of a relationship between 'absolute consciousness' and temporality is made possible by the hegemony of the time of the discrete 'now', posited as distinct from the past and future. The inconspicuous familiarity of the mathematical image of a line falsifies the *truth* of our lived temporality. The 'numerical identity' of a transcendental apperception is for Heidegger incapable of expressing the 'unity' of lived existence. 'Consciousness', in this way, is a discursive falsification that suppresses the temporal existence of the questioner. In this falsification, an *eigentlich* self-interpretation of existence is displaced, and replaced, by a public-understanding of an a-temporal 'absolute consciousness'. For Husserl, as we will see later in Heidegger's criticism of the phenomenological and eidetic reductions, there is a move far away from lived experiences towards an obliteration of this facticity in a derivation of an 'image' of a linear, temporal order.

In a way similar to Hegel in his *Phenomenology of Spirit*, Husserl begins with the flow of lived experiences – the immediacy of sense-certainty. There is a temporal flux of discrete moments shading off. Amid such shading, however, it is fathomed that there persists a limit-ing of receding difference which is constant: it is here where Husserl finds the thread for his suppression, reduction of original temporality and the construction of objectivity and identity. This latter is founded upon difference, yet as a recurring difference which thus indicates the identical. Husserl gives this 'identity' a specific meaning vis-a-vis his theory of 'absolute consciousness', and restricts any investigation into the being of a temporal objectivity within the horizons of an extratemporal possibility. The generic time of 'absolute consciousness', itself the a-temporal source for the 'objectivity of time', is the coxswain that beats out *the* rhythm of public existence and its lived experiences. Husserl reveals a susceptibility, however, to a *destruktion* of this 'absolute position of consciousness' which he exposes in his telling statement:

Theoretically, this process is to be thought of as capable of being continued without limit, although in practise actual memory soon breaks down.[59]

Husserl very clearly admits that he is concerned with a theoretical interpretation of time, of objective temporal positions, each with a strict identity, and not with the temporality of lived experiences or of lived existence, for such lived existence would eventually 'break down'. Original temporality is that which breaks down, or breaks in, amid this playspace of temporalizing, of existence. Heidegger locates his sense of *theorein* (θεορειν) in such a breach, where an existent thinking breaks down, becomes bored, distracted, anxious amid be-ing. Radical finitude is uncovered in the breakdown or breakthrough, in the transition amid an unfamiliarity to the *place* of the strange, uncanny. Significantly, this radical mutability of finite existence is one that must include the existence of the philosopher, the questioner. It is only relative to existence that *logos* has any meaning at all. Husserl's notion of 'lived experience' is radically distinct from Heidegger's indication of lived existence. The former, conceived as an immanent happening of consciousness, posits the strict identity of objective temporal order, and, with respect to its objective intention, becomes a reservoir of data for the 'reflections' of a ' pure consciousness'.

This emphasis upon strict objective identity, fixed in a linear temporal order, is re-enforced by Husserl in his rejection of the ambiguity of his teacher Brentano's idea of 'primordial associations'. This latter indication suggests that Husserl's notion of primary remembrance, as THE PAST, would, from the perspective of a 'primordial associations', undergo continuous modification, *bleeding* into the present, coalescing with 'presence' as still having-been. Such an indication foreshadows Heidegger's reply to Husserl with respect to the significance of his phenomenology of ecstatic temporality. Husserl rejects Brentano's notion of 'primordial associations' as absurd in that it would violate his criterion of strict identity, that the 'past is past', or, that after the peak of a primary impression (the source of being) has 'run off', withdrawn into a 'primary remembrance', and hence into a 'past'. All that is now possible with respect to it is a recollective or reproductive experience of such a past, as an image, a copy, as with a shade in Hades, which is in diminished similarity[60] with respect to its lived existence.

This dispute is reminiscent of statements by Parmenides in Plato's dialogue, a work of some concern to Heidegger, where Plato speaks of possibilities of an existence where the 'One' (**hen**) interferes 'in' this world of Many, and of the distortions which ensue, in almost comic detail. Between the two extremes of a One as such and an in-existence of said One, between silence and noise, there is a middle world in which the One comes into a tenuous harmony with the Many, or in other words, with predication, categories, expression, voice. Yet, given the precise character of the terms, joined in this harmony, this middle-world is a playspace which, from a perspective of strict identity, would be absurd. In other words, for Husserl, following Zeno, existence, a creature of primordial associations, *is* absurd – it has 'impossible attributes'. Yet, existence **is**. Heidegger was closer to Brentano than Husserl in this dispute, but in the end, he draws a differing conclusion with respect to the relation between the 'one and the many'. Since being is not a real/ideal predicate, it is neither subject to the criteria of logical identity, nor to the law of contradiction. Moreover, since original temporality is the horizon for the question of being, an original temporality is the precondition for those very things of which there is predication. Ecstatic temporality is not adverse to a possibility of impossibility, to the absurd, since these latter judgements have jurisdiction only within the horizons of real/ideal predication, and its totem 'common time'. Heidegger projects an indication of temporal lived existence, in which the past is an ecstasis of temporality which not only erupts into awareness as a breach, but exists not only in terms of historicity (as it is relative to Dasein in its own radical temporality), but also with respect to 'events' of being-in-the-world. Such an 'absurdity' suggests that a past could be re-lived, re-activated, perhaps with even more power than its original.

Such a possibility is due to his transcendental imagination, and his identification of this power with original temporality. For Husserl, imagination is always reproductive, and thus, always deficient in its power vis-a-vis the 'original'. Phantasy in this way is always distinguished from perceptive 'reality'.[61] Heidegger, set free from the criteria of real predication or logical identity, has described what, in Husserl's terms, could be designated as a 'primordial phantasy' in his indication of ecstatic temporality. Heidegger contends that Husserl's severance between immanence and the transcendent and his overriding concern for strict identity forces him to deny possibilities that will not fit into his framework. As with his teacher Brentano, he denies the 'many' for a One. Heidegger, on the contrary, evokes his 'primordial phantasy' of a 'middle world' of impossible attributes, a *playspace* for the arrival of entities into the world.

Chapter 2

An Indigenous Conceptuality of Dasein

Whenever a phenomenological concept is drawn from primordial sources, there is a possibility that it may degenerate if communicated in the form of an assertion. It gets understood in an empty way and is thus passed on, losing its indigenous character, and becomes a free-floating thesis.[62]

Indigenous expression and the being of the self

From his first published writings on Duns Scotus in 1915 and his indication of 'haecceitas', of *this-ness*, we glimpse Heidegger's determined attempt to leave a phenomenology dominated by theory, by 'unworlded' abstractions, and to gain a 'toehold', a way of access to and expression of the temporal sense of 'matters themselves'. For the next ten years, Heidegger unfolded an 'indigenous conceptuality' of being-in-the-world. Heidegger understands his project, in an echo of Nietzsche's manifesto of art as a *counter-movement to nihilism*, as a counter-ruination *amid* theoretical falsifications of lived existence.[63] Heidegger undertakes to dismantle the regime of falsification through a *destruktive* appropriation of the theory at hand in light of the question of the source of its expression, of its 'conceptuality'. Heidegger traces the expression of phenomenology to a temporality of ruin (**Ruinanz**) and sets out from this facticity in his own attempt to articulate an indigenous expression of existence. Turning away from the protocols of Neo-Kantianism, Husserl's phenomenology and logical positivism, he points to that which is 'there', to the phenomenon in a pre-theoretical beholding of 'truth'. The intimacy of this disclosure is not that of Heidegger's 'shattered'[64] Platonic-Christian *naivity before God*, but is the openness of a situation of self-interpretation amidst the playspace of disclosure and concealment.

The being-there of the self, of the one amid its world, is, as *freedom*, outside of common, or more precisely, *generic* time, which cannot express a *singular* sense of existence. Amid this pre-understanding of one's own predicament, we can take a step back towards 'primordial sources' for a self-expression of our own original temporality. Self-expression, in the context of the question, gives access to, uncovers, sets free, original 'matters themselves'. It intimates an original *topos* of 'concept formation' amid, as Krell writes, the 'raptures' of ecstatic temporality.[65]

Within a logically disciplined 'system of consciousness', there can be no encounter with *this-ness*, such a possibility cannot even be articulated, for this would violate the circuitous discursivity of necessary and universal norms of thought, of strict knowledge against this desire to behold the 'things themselves'. With clear Platonic overtones, Kant asserts that the distance (and authority) which Reason maintains with respect to the 'dimensia' of sensibility necessitates a procedure of subsumptive predication in a regime of theoretical truth. One can only be perplexed over the possibility of a *beyond* of this familiar web of 'consciousness', as we can never *know* anything regarding this 'outside'. In this way, if we accept these limits, the question of the 'intimacy' of a pre-understanding, and therefore, of a radical phenomonology of original temporality is rendered mute (or shifted into the realm of practical reason). From the turrets of Leibniz, Kant and Husserl, intimacy is quaranteened into those makeshift domains of private languages, poetry and madness, such as was done with James Joyce's *Finnegan's Wake*[66] (1928), 'silent' *de facto* in the sense of the logical proposition. 'Consciousness' is suspicious of discourses which fall away from the rhetorics of logical formalization, routine clarity and generic precision.

Heidegger contends that phenomenology must be a vital response to this 'crisis', one which will allow us to break through our 'rote' procedures, and to re-aquaint ourselves with our be-ing after the fall, the catastrophe of the breach, portrayed in Kisiel's discussion of Heidegger and the 'Great War'. Yet, how can philosophy express that which immediately bestows itself amid one's own world, how can we think for ourselves, as the point of departure of our expression is that of the discourse of historicity? How do we access/express *things themselves*, not merely parroting a historical philosophy, but apprehending and speaking about such things in our own voice? With which words can we 'express' the ecstasy of existence? Are we not thrown back into the same dilemma as Kant in his futile attempt to deduce the ground of the legitimacy for the application of received concepts to the manifold of appearance from the (violent) spontaneity of apperception? Since the concepts he traded were inadequate to the expression of the intimacy of lived existence, their continued application, for Heidegger, is merely a covering up of the truth of existence with a pre-fabricated expression. How can discursivity approach **this** – memory, hope, awe, belief – without destroying it? Heidegger articulated his radical phenomenology of ecstatic temporary within the horizon of these questions. The method of formal indication was characterized as a dedicated submission to the phenomenon, a finite knowing, or, in the words of *Kant and the Problem of Metaphysics*, a pure, *sensible* reason. Such a paradoxical thinking is characterized by receptivity, an openness to that which is there, listening, amidst 'no-thing'.

Indigenous expression is articulated amidst finite, lived existence. It cannot account for its own legitimacy, but simply discloses that which is there. But have 'we' already said too much? Must we not heed the early Wittgenstein's warning to pass over that about which we cannot speak in silence? Is there an expression beyond a methodological silence?[67] In the following pages, I will explore the character of radical phenomenology in the context of Heidegger's engagement with the phenomenology of Husserl. It will be out of this *destruktion* that Heidegger will unfold the basic features of his analytic of existence which would become his 'Sein und Zeit' project.

The birth of phenomenology

Turning to the 1925 lecture course, *History of the Concept of Time*, we will trace Heidegger's projection of the origins of the 'phenomenological movement' against the background of its emergence as a response to a 'crisis' in philosophy as such. In the self-understanding of the early phenomenological movement, a return to the 'matter themselves' was regarded as an explicit resistance to the degeneration of thought to a mere appendage to a logico-scientific 'machine'. With the collapse of idealistic systems amid the convulsions of the nineteenth and twentieth centuries, there arose 'positivist' ideologies of an 'arid and crude materialism', and 'philological criticism', filling the vacuum that had erupted with the 'analytical fragmentation' of a 'whole man', cut up (**ratio**), classified, divided against itself.

Amid this 'crisis', there were differing responses. Dilthey, Heidegger relates, had called for an independent method for the human, historical sciences. He sought to claw his way out of the vortex of a hegemonic 'moralist' sociology. With the intimation of the specific character of 'life', Dilthey opposed the application of the methodologies of the 'natural sciences' to human relations, advocated by Mill and 'revolutionaries', such as Engels, who manufactured his own 'scientific socialism'. Dilthey advocated an expression of an indigenous sense of **life**. Heidegger describes Dilthey's project:

> Its task is rather to regard 'life' itself in its structures, as the basic reality of history. The decisive element in Dilthey's inquiry is not the theory of the sciences of history but the tendency to bring the reality of the historical into view and to make clear from this the manner and possibility of its interpretation.[68]

Heidegger contends that this vision of 'life', conceived as an intimate disclosure of the 'ways and byways' of phenomena, intimates a *topos* (**Da**) 'beyond' the blindness and violence of objectified 'consciousness'. As is well known, it was his affinity with the *life-philosophy* of Dilthey and to Nietzsche which led to the great divorce within twentieth-century philosophy between the logical positivism of Carnap, Ayer et al. and the phenomenological movement which traces its genealogy through Husserl to Brentano.[69]

Brentano, Husserl's teacher, had resolved that philosophy must 'draw its concepts from its own matter',[70] to simply clarify 'what' is given, the phenomena themselves without 'constructions'. Brentano admitted however that phenomena must be 'classified according to basic structures', an ordering which is 'always done from a point of view'.[71] Not mere imaginative orders, contexts of objective, physio-logical 'relations', but perspective, a way of seeing 'drawn from the actual elements themselves'.[72] Brentano's descriptive psychology set out the parameters for access in that 'classification can be made from a prior familiarity with the objects . . .'[73] by which is meant, in the context of Heidegger's project, ways of being, pre-theoretical, pre-practical orientations of existence. Brentano, prefiguring Heidegger's *destruktion* in the Kantbook, asked the question of the relation between psychic phenomena and what is designated as the 'physical'. He answered that these latter 'stems' can be traced to a common rooting in an 'indwelling' of 'something objective', in the 'structure of action'. Brentano designates this indwelling as 'intentional inexistence',[74] which is conceived as the ground of epistemic relations, such as representing, judging or willing, conceived as lived experiences.

The notion of *intentionality*, to which Brentano's conception is related, traces its genealogy to the Scholastic *intentio*, a 'directing itself toward' and to the *intentum*, that toward-which there is direction. For Brentano, intentionality is a self-directedness that is grounded in representation, among which is included the *phenomena* of representation, such as judgement, interest and emotions. Representation is thus considered to be a 'basic' comportment of the original unity of intentionality. Heidegger objects to this characterization of intentionality in that representation, even if broadly conceived, remains primary to the *things themselves* due to a starting point in the severed stems of the psychical and physical. In this way, following Aristotle, 'intentional inexistence' is a representation that mediates a relation of opposing representations. Husserl, for his part, and which we will see in more detail below, answered the question strategy of his teacher through an elimination of the dogmatic conception of the transcendent object. In both cases, however, Heidegger contends that there is a suppression of the matters, of things themselves. Indeed, the starting point in severance forbids *de facto* any access to the matters themselves. Heidegger lays out a provisional exposition of intentionality as a 'structure of lived experience', as the 'comporting of all relations of life'. He contrasts this, following his rejection of the primacy of representation in Brentano, with a similar strategem in Rickert's Neo-Kantian 'philosophy of immediacy' by pointing out that there must be a way of access to a disclosure of 'matters themselves'. He admits that he cannot avoid construction, but seeks instead one which is a preparation for a formal, or in the case of *Being and Time*, an 'existential' indication through which phenomena are released so as to 'speak for themselves'. Despite his basic agreement with Rickert's criticism of Husserl's 'eideticistic' reduction, Heidegger criticizes the former for not allowing his immediacy to be immediate. Rickert's may speak of immediacy, Heidegger suspects, but he has simply repeated the architectonic of 'consciousness'; he cuts, distributes 'being' into a *psychic* and *physic*, immediacy over against the mediate. Amid 'consciousness', immediacy becomes 'semblance', shades. Rickert thus remains trapped in representation, covering over matters themselves. For Heidegger, one of the most profound errors of the reaction to the hegemony of the 'Scientific Worldview' was a return to the Kantian philosophy in the guise of 'Neo-Kantianism'. In other words, it was a 'renewal' which took place not as 'an original return to the matters at issue, but by going back to a historically established philosophy . . .'[75] Any 'authentic', honest, phenomenology, on the contrary, must return to the 'matters at issue'. We will thus turn to Heidegger's account of how the early phenomenological impulse sought and failed to overcome the barriers to the things themselves erected by Neo-Kantianism, to find Ariadne's thread out of the labyrinth of representational 'consciousness'.

Breakthrough discoveries of phenomenology

Heidegger's account of the early phenomenological movement is not an 'objective' history, but a preparatory investigation which will lead not only to his *destruktive* criticisms of Husserl's captivation by the mythology of consciousness, but also his own appropriation of the discoveries of the movement. In what will become the first half of the lecture course, Heidegger will meditate upon and challenge the articulation of phenomenology in the Sixth of Husserl's *Logical Investigations*.[76] It will be through this necessary detour that Heidegger will arrive at his own indigenous 'conceptuality' of the *existentials* in his phenomenology of ecstatic temporality.

There were three 'breakthrough' discoveries of early phenomenology according to Heidegger: intentionality, categorial intuition and an original sense of the *a priori*. It will be these innovations that he will deploy to overcome the retrogression into Neo-Kantianism.

Intentionality

Heidegger seeks to bring phenomenology back to comportments with 'matters themselves'. This entails, as we have already seen, a challenge to the image of representational intentionality as a co-ordinating structure for the severed stems of the psychic and physical, where any psychic act has its counterpart in the physical. This account of intentionality is countered by Heidegger with a prominent and decisive example: that of a hallucination of an automobile sailing over and across the lecture room, where the linkage between the psychic and physical is disrupted. Yet, if all there is is representation, how do we know there is a disruption? Heidegger is concerned with *comportments*, ways of being amid the 'matters themselves', which have the structure of a directing towards, which even a hallucination exhibits. He criticizes Rickert for his compartmentalization of intentionality into the immanences of judgement, value, 'consciousness', and thus for his failure to apprehend things themselves in their immediacy. With his own sense of intentionality as the 'comporting of all relations of life', Heidegger, anticipating the post-structuralists, contends that since intentionality has been severed from representation, there need no longer be question of a co-ordination of the stems of the psychic and physical. Indeed, there is no necessity of remaining within a philosophy of 'consciousness' at all in that phenomenology is a 'not knowing' in Kant's sense. Heidegger challenges Rickert's concept of 'immediacy' with being blind to intentionality as a comportment with an entity itself. Rickert is captivated by a 'box'[77] theory of 'consciousness' that necessitates he be blind to phenomena, which are themselves displaced in favour of a *theory* of immediacy. Heidegger insists that we must witness intentionality as a *directing itself towards* in its singularity, as lived existence. He declares: 'All theories about the psychic, consciousness, person, and the like must be held in abeyance.'[78] The perceived entity shows itself from itself as primary – names, assertions – are brought to the 'thing'. Heidegger is concerned with simple apprehension, letting something be seen, *before* assertion (*logos*).

> In opposition to this scientific account, what we want is precisely naivete, pure naivete, which in the first instance and in actuality sees the chair.[79]

This seeing is pre-theoretical, pre-practical, which is possible with regard to Heidegger's specific *epoché*, in which he brackets epistemology, psychology and 'theories of perception'. One does not bring a theory or the expression of the thing from without in the manner of Kant. For Heidegger, 'naive' characteristics are 'read off' a thing itself via 'a simple envisaging of structures'.[80] This reading unfolds the perceived in the strict sense as a phenomenon *showing itself*, before the artifices of representation.

The 'mythology of consciousness' has instigated, Heidegger states, a severe confusion with its characterization of perception as a subjective picturing of an objective outside. Such a 'theory' always leads to an infinite regress, and of the many language games which are designed to avoid such a regress. Sherover in his remarkable work, *Heidegger, Kant and Time*, contends that, in that Heidegger does not begin with the stems, the projection of a common structural rootedness of the stems in his interpretation of Kant (as we shall see) obviates this infinite regress. For Heidegger, however, the logical problem of the mythology of consciousness is not what is at bottom the most problematic about this mythology. Instead, it does not correspond to the 'simple phenomenological findings'.[81] It is un-phenomenological.

With perception, Heidegger describes, the entity is 'there' in its 'thing whole'. The sameness of the entity persists amidst adumbrations, shadings, which are manifest via a phronetic looking around, as circumspection. In other words, a *perceived* is *imagined* to be the same, and hence, its 'unity' is primary to its aspects. In this way, *perceivedness* as lived intentionality, as the intentionality of lived experience, is primary, is a 'whole', as with Dilthey's holistic *a priori*. The 'unity' of the *intentio* (the why something is intended) and the *intentum* (the how of being intended) must be understood from out of the 'unity' of intentionality itself. An *intentum* belongs to every *intentio*, Heidegger states, with a 'specific structural interrelation'.[82] Each intention is a projection upon a possible 'fulfill-ment'. Intentionality 'is' the *place* of all acts, the formation of concepts from raw meaning. It is not concerned with Rickert's 'subsequent coordination' of stems, but is an original unity, a 'belonging together of intentio and intentum'.[83]

Heidegger surmises that his innovation in the grammar of intentionality poses difficulties for the phenomenologies of Husserl and Scheler, who assimi-late the structure of intentionality, via Rickert and Brentano, into a blind nomenclature of consciousness, reason and spirit. Heidegger calls for a distinct radicalization of phenomenology. He foreshadows:

> intentionality is not an ultimate explanation of the psychic but an initial approach toward overcoming the uncritical application of traditionally defined realities as the psychic, consciousness, continuity of lived experience, reason.[84]

Categorial intuition

In light of his bracketing of 'traditional' protocols and standards of conceptual-ization, Heidegger indicates an alternative mode of conceptuality in *categorial intuition*. Intuition is the simple apprehension of the bodily given as its shows itself. However, there is also simple apprehension of the *categorial* amid the eve-ryday pre-understanding. In this way, categorial intuition is the self-expression of 'intentionality'. With Heidegger's way of thinking, original perception 'gives the demonstration'[85] for empty intention. He contrasts this fulfilment, as a 'structure of evidence', with that of 'feeling of evidence' of Rickert. This feeling is a 'psychic' process which is necessitated in that a *said* 'inside' has no direct experience of an 'outside'. Heidegger dismisses this theory as not correspond-ing to 'findings'. 'Evidence' breaks in as the disclosure of a sense of being that is accentuated from this 'originarily intuited matter',[86] a breach amid this situa-tion of inconspicuous phenomena. This 'breach' throws the familiar into conspicuousness. Heidegger insists that phenomenology must break with a notion of truth abiding only in judgement. He refers to the 'Greek' association of being with being-actual, as 'non-relational single rayed acts', which is an emphasis on 'matters themselves'. With this contrast, as was indicated in the previous chapter, the sense of 'truth' is provisionally widened.[87] In his *destruk-tion* of the Sixth *Logical Investigation*, Heidegger explores the distinction between the simple as a *single level act* and those acts of expression that entail *multi-level acts*. Removing himself from the severed stems, he exposes the alleged *naivity* of the simple level act as 'pervaded by categorial intuition' and exhibiting a 'high degree of complexity in its act-structure'.[88] He describes this complexity of acts through a distinction between two types of acts. On the one hand, there is the 'unity' of simple lived intentionality which is of a 'single level character' – single-level acts. On the other hand, multi-level acts are those in which there is *expression* of the 'matters themselves'. There is, moreover, an inherent relation betwixt the two 'objectivities', as one is built upon the other. The simple act is repeated via the categorial act, in a new objectivity. The multi-levelled act is an 'act of expression'. It is a 'founded act', assertion, *logos*, and 'discloses the simply given objects . . . precisely in what they are'.[89]

The simple is disclosed through expression, 'precisely in what it is', in an accentuation of a *topos*. This pointing out is a reciprocal, dual-directional relation, a 'double direction' belonging to this being of states between two differing acts. The new objectivity is expressed, it expresses itself via its discourse, as putting together (synthesis) and taking apart (diairesis) which taken together give forth an 'objectivity' as an accentuated 'state of affairs'. This is *logos, founded* upon simple apprehension. This expression is an indication, a *naming* through which phenomena are accentuated. Simple apprehension is however always already pervaded via categorial relations, and thus a new objectivity will not only express the truth of an entity, but an entity itself is only fathomable on the basis of such a pre-understanding. Yet, simplicity is primal in its figurative giving of intuitive 'unity', a simple, self-contained, *given* amid a 'glance of the eye'. 'Matters themselves', and not assertions, remain the exemplar of an original sense of an *a priori*.[90]

Ideation, on the contrary, gives a new objectivity founded upon a fundament of individuality, but it purports not to mix with this fundament. 'Ideation', as narrated in the 'mythology of consciousness', is an intuition of the universal. This categorial act gives itself its own object, idea (ειδος), as outward appearance, the 'look' of 'something' – its 'what'. The act gives a 'universal of individuations', an ideal 'unity' towards which one 'looks' in a concrete act of comparison. However, since simple apprehension is already always pervaded by categorial acts, there are no simple acts 'versus' complex acts, and thus, for Heidegger, there could be no 'universals' which would not always be able to trace their lineage to the matters themselves. Categorial acts, for Heidegger, must be founded upon simple apprehension, in whatever manner this takes place and upon whatever 'ground'. They do not 'float freely'. Heidegger denies to ideation its escape from these 'matters themselves', contending that, as Aristotle wrote, 'the soul never thinks without an image'. Thought comes to be disclosed as an accentuation of 'something' already given; finite thinking is 'grounded' upon a founding *sensuousness*. Indicating the beginnings of his *destruktion* of phenomenology, Heidegger contends that Husserlian 'acts of ideation' do not 'intend' or 'relate to' sensuousness, but nullify it, and, as pure categorial intuitions of generality, must therefore be *put out of play*. In its denial of the *transcendens* in favour of a reduced immanence, the 'ideas' do not follow along with the movement of the phenomena in the truth of its 'mixed' character.

An original sense of the *a priori*

The third breakthrough discovery of phenomenology is its indication of an 'original sense of the *a priori*'. For Heidegger, the *a priori* presupposes an understanding of time, and in this way, he lays out its provisional indication in formal terms, as the 'earlier'. He takes to task any concept of an *a priori* as the "knowledge of the subject" – '. . . before it oversteps the bounds of its immanence.'[91] On the contrary, he claims, the *a priori* has 'nothing at all to do with subjectivity'.[92] As we have seen vis-a-vis categorial intuition, 'ideas' are 'read off' the 'things themselves', 'structures' come via a careful accentuation of the 'matters themselves'. In this way, the *a priori* is 'structurally earlier', and thus, it is a 'title for being'. *A priori* truth must, as Heidegger describes, be 'universal', 'indifferent' to subjectivity, and be directly accessible in simple and originary intuitions. *A priori* truth, in this way, is not an aspect of *knowing*, of statement and reason (*logos*), but is the horizon of the situation, *topos*, of existence, which is disclosed amidst a self-interpretation of Dasein.

The maxim 'to the matters themselves' is regarded, by Heidegger, as the announcement of a movement in philosophy which sought truth with minimal construction and without the trading of free-floating and groundless concepts. The maxim invites us to demonstrate *this* from a 'native ground', to 'secure' the matters, through an openness to the phenomenon. Heidegger states that a *living* philosophical 'logic' is a thinking which abides to the 'lawfulness of the object'.[93] Thinking is, at the same time, self-expression, as a 'meaningful fixation of what is thought'.[94] In this light, intentionality is to be considered in its *a priori*, as an original *topos* of the phenomenon. This modality of treatment of the 'field' is a 'simple originary apprehension',[95] a formal indication of the phenomenon, a self-expression of phenomena amid one's own 'direct self apprehension'. In this way, as Heidegger concludes, finite self 'knowing' is an 'analytic description of intentionality in its *a priori*'.[96]

Heidegger turns to the event by which the idea or concept, the assertion, became free-floating, as it had severed itself away from its tutelage to the phenomenon, just as Apollo severed himself from Delos/Dionysus. To this end, he outlines in his lecture two historical senses of the term phenomenon 1) the manifest, a non-referential, originary self-showing of entities themselves; 2) semblance or appearance which is a standing in, a reference to that which conceals itself in being announced, such as the Kantian thing-in-itself in the relation of phenomena and noumena. It is pointed out by Heidegger that the latter sense is a modification of the former, and relies upon an original, 'Greek' sense of phenomena since even an appearance or a symptom is a self-showing, despite its referential function. The word *logos* (λογος), traced from *legein* (λεγειν), has the sense of a making manifest via discourse, *apophansis*, is a letting something be seen from itself, a saying that is drawn from the subject matter, from what is being talked about. Concretely, *logos* is voice or a reticence, an indication or 'pointing out and letting something be seen.'[97] *Logos* is thus, in its basic state, semantic, a general signifying, or 'something vocal which shows something'.[98] Apophantic discourse, however, in a post-Kantian sense, has become restricted discourse, it points out the 'spoken' to be seen 'as' itself, in theoretical *logos*. It is with theoretical 'consciousness', Heidegger contends, that the first sense of phenomenon transforms into the second. Semantic discourse, co-opted into theoretical statement, can no longer give a place for expression of a non-apophantical voice in 'an exclamation, a request, a wish, a prayer'.[99] The phenomenon can no longer express itself, but is represented and positioned in theoretical statements which orchestrate a specific limitation of semantic discourse, where *logos* takes on the meaning of *theorein*.

The meaning or sense of phenomenology, for Heidegger, however, takes its cue from an 'intrinsic and material relation' between *logos* and phenomenon, 'letting the manifest in itself be seen from itself'.[100] That which is manifest is not material content, entities, but the finite *how* with respect to an encounter with this being (**sein**) of entities. This *how* is a 'way of encountering something', a pointing something out, 'how' matters must be prepared, to be there for interpretation and expression, indication, 'laying open and letting be seen'.[101] In preparation for his *destruktion* of Husserl's phenomenology, and hinting at broader claims with respect to the historical 'forgetting' of the sense of being, Heidegger writes that which is seen, the truth as *a-lethea*, can be suppressed, covered up, concealed. 'Being covered up is the counter-concept of phenomenon.'[102] Being-covered-up is falsity, undiscovered, buried, disguised. The shoots of falsity are assertions, which disengage from the phenomena, lose their rooting in the soil of finite existence. In their remoteness from 'matters themselves', voice is 'lost to itself' in a 'detour' of assertion. Phenomenology becomes 'hardened in its results'.[103] Heidegger insists that phenomenology must be 'critical of itself in a positive way'.[104] This *destruktion* is an un-covering of phenomena that, in its desire for truth, sets forth what he calls a 'picture-book phenomenology'.[105] Categorial intuition is 'founded' in that it requires that there be an 'intuition of an example'. Intentionality, in this way, must be read off from an 'exemplary ground of the field of concrete individuations of lived experiences'. The original sense of the *a priori*, the ways and byways of this ground, requires the accentuation of phenomena. From this 'field', there is disclosed a 'character and type of being of this region'.[106] Yet, contrary to Husserl, Heidegger insists the character and type of being in this region is one of temporality and the *topos* is that of finite existence, the 'Da' of Dasein – and not 'consciousness'.

Husserl's thematic field of 'pure consciousness'

The aim of Husserlian phenomenology is to 'distill out' a pure essence of 'consciousness', of *pure* lived experiences, a pure 'ego' from a point of departure in the 'lived experiences' of this 'ego' in its 'natural attitude'. The latter is an individual 'stream' of lived experiences, of the human subject as a 'real object in the natural world'. The act of self-directedness towards our own experience is 'reflection'. We turn our gaze upon our own acts. Reflection and that upon which is reflected belong to the same sphere of being that Husserl describes as 'immanence', a sphere 'apart from any and all {properly} essential unity with the thing'.[107] There is a 'gulf' between immanence and transcendent perceptions of things. This stream of 'pure consciousness' is a 'self-contained totality' which excludes 'every thing', 'every real object, beginning with the entire material world'.[108]

At the same time, however, 'consciousness' is also regarded as 'present' factually 'in' physical *things*, in that it has a 'never-absent physical connotation'[109] which is merely a re-statement of the point of departure. Heidegger writes that in order to isolate the pure essence of 'consciousness', amidst its 'double involvement', Husserl contends that we must not live in the perception, in the apprehension of the thing, but as intentionality, we must 'live thematically in the apprehension of the perceptual act and of what is perceived in it'.[110] This abstinance from a transcendent world symptomatizes Husserl's *epoché*, his thematization, construction, of an 'attitude of the immanent reflective apprehension'. This is a not-going-along from an 'act' towards things, but staying with an 'act' itself.

For Husserl, it is possible to enact a phenomenological suspension across this whole range of acts via a reduction which, Heidegger describes, lays out the sphere of acts and its objects in the 'uniformity of a specific sphere'.[111] The field is a unique singularity, which is *my* stream of 'consciousness'. This singularity, however, in its concrete acts, is subjected to another *eidetic* reduction, which suspends all individuation to reveal the structure of the pure field of 'consciousness'. In this reduced field, an object of immanent perception is '**absolutely given**'.[112] This 'stream' is a region of 'absolute position', and in reference to 'Kant's thesis about being', 'pure consciousness' is the 'sphere of absolute being'.[113] With this determination of the meaning of being, Heidegger suggests that phenomenological reflection reaches a 'climax', an 'end', or, as Bataille might insert, a 'little death'. He asks pointedly before undertaking his *destruktion*, how this 'thesis about being', especially as expressed in the eidetic reduction, can relate to a 'unity of the real human being?'[114]

Heidegger's *Destruktion* of phenomenology

In a climax to his only sustained criticism, Heidegger questions Husserl's fourfold determination of 'pure consciousness' as 1) immanent being; 2) absolute being absolutely given; 3) as independent of reality; and 4) as 'pure being'. Heidegger asks *how* 'pure consciousness' is to be rooted in the 'matters themselves'? In all four cases, there is neglect of a question of the sense of being of the entities or acts, as for instance in the neglect of the *intentum*. In the first case, that of pure consciousness as immanent being, the being of immanent 'acts' is left undetermined. In the second, the being of this alleged 'absolute' is left in silence. In the third, 'pure consciousness' as that which is independent of reality is revealed as an 'earlier' that survives an 'annihilation of the world of things'. In the fourth, the meaning given to being is that of an ideal, not an 'actual' being. Heidegger declares:

> All four determinations of the being of the phenomenological region: imma-
> nent being, absolute being in the sense of absolute givenness, absolute being
> in the sense of the *a priori* in constitution, and pure being are in no way drawn
> from the entity itself.[115]

Heidegger's indication of an original sense of the *a priori* seeks to demonstrate that Husserl has 'ruined' his initial breakthrough with a premature climax, in his answering too soon, naming a **what** – the being of the field of intentionality as 'pure consciousness' and all that such an answer entails. The thing, this being of the entity, as a field of intentional acts, is covered up by traditional theories of 'consciousness'. In this light, Heidegger contends,

> the elaboration of pure consciousness as the thematic field of phenomen-
> ology *is not derived phenomenologically by going back to the matters themselves* but
> by going back to a traditional idea of philosophy.[116]

In this way, the matters which are still in need of determination have been dis-
placed by the insertion of a traditional field which has embedded in itself a logic *of its own*, alien to the matters. Indeed, Heidegger contends that it is Husserl's procedure of eidetic reduction which erases the matters, in the ideali-
zation of singular being into mere *res extensa* in the manner of Descartes. What is excluded in the reduction is precisely this factical, ontic ground which founds the possibility of phenomenology. In other words, in the construction *via nega-
tiva* of an *a priori* in 'pure consciousness', Husserl has come into conflict with this original, 'Greek' sense of an *a priori* in that he forsakes a phenomenological description of the 'that' of a phenomenon and the 'how' of the encounter. *Stepping not beyond* the *topos* of the 'natural attitude', of naivety, Heidegger points out, as with the exception of this example of hallucination which disrupts a regime of epistemological discipline, that this ecstatic be-ing evades eidetic reduction. He describes the predicament of such a be-ing:

> But if there were an entity whose *what* is precisely to be and nothing but to be,
> then this ideative regard of such an entity would be the most fundamental of
> misunderstandings.[117]

Since there is such an entity that is nothing else but *to be*, it is 'phenomenology' that betrays a fundamental misunderstanding. Heidegger contends that it is Husserl who has himself implicitly asked and set out an answer to the question of being: Being, the sense of actuality of something, is determined as a being for 'consciousness'. Being is a *manifesting* in, and for, 'consciousness'. Heidegger claims that not only is such a formulation premature, but that this definition harbours 'hidden' ontological implications and tacit commitments which *taint* Husserl's entire phenomenological endeavour, beginning with his determination of the sense of 'natural attitude'.

Husserl determines the 'natural' in the way of 'natural science' – the human being is conceived as a living being, a zoological object. Heidegger claims however that this is indeed an 'unnatural' starting point: it is a theoretical position, it is an attitude. Indeed, the 'natural attitude' is merely the 'other side of the same coin' of the reduced 'pure consciousness' – it begs the question of a 'pure consciousness'. The meaning of the being of acts is, in this way, 'theoretically and dogmatically' defined in advance as the meaning of being of Nature. Husserl's sequence of reductions then is a de-naturalization. In this way, Heidegger, keeping in mind the indefinite yet indicative original sense of the *a priori*, claims that Husserl does in fact give an *answer* to the 'question of being'. Yet, the answer and question remain undiscussed, buried, silenced by his 'mythology of consciousness'. He pretends not to be concerned with the question, yet, this semblance of indifference betrays a hidden nexus of 'binding commitments'.

For Heidegger, the question of being is provoked by a conflict between the divergent senses of the *a priori*: between the 'entity' and 'pure consciousness'. This question is necessary since, in his final determination of the 'most radical of all distinctions of being', that between 'pure consciousness' and of a transcendent, Husserl, via his *eideticist* strategy, has neglected/obliterated one of the poles of distinction, this specific be-ing of comportments with the transcendent. There is a neglect of this be-ing of 'acts' and of the sense of being as such, a question which does not float in freely from the air, but is provoked via Husserl's own repetition of what, for Heidegger, precisely falsifies the phenomenon.

It is in light of such a technicist repetition that Heidegger deconstructs Husserl's 'supplement' of personhood, of the *spiritual lifeworld* in *Ideas II*, in its attempt to give a breathing space for 'personality' with respect to a universalistic 'scientific naturalism', much the same way as Kant's limitation of reason to 'make room for faith'. Husserl's supplement is, as Gadamer would concur, 'ontologically the same', since these 'constituted' distinctions remain dependent upon the already established *a priori* ground of *Ideas I*, upon 'pure consciousness', the universal structure of reason. Indeed, and in a way similar to the falsification of the being of the natural via an idea of a 'natural attitude', Heidegger insists 'personality' as a 'constituted' region and theoretical 'normalization', with all that it implies, covers up this being of the 'acts'. The question of the being of 'personality' is not asked. In this way, this supplement repeats the suppression of acts in the sequence of reductions.[118]

Heidegger credits Scheler with laying new ground in raising the question of the being of the acts. This question strategy, in tune with a bracketing of 'consciousness', rejects the idea of the psychic as an interpretation of the being of a 'whole person'. Heidegger quotes Scheler:

> The sole and exclusive mode of its givenness {of the person} is rather its very performance of its acts (including the performance of its reflection upon its acts) – in living its performance it simultaneously vitally experiences itself.[119]

For Heidegger, this is only a first step, however, in the 'twilight of the idols' of a deconstruction of the rhetoric of 'consciousness'. It is only a beginning since silence reigns regarding the question of the mode of being of an 'act performance' and of the 'performer' – of the questioner. Heidegger charges that phenomenology fails to live up to its initial breakthroughs. Its assimilation into the tradition prevents it from making an 'original leap to the entity'. He declares:

> Not only is the being of the intentional, hence the being of a particular entity, left undiscussed, but categorially primal separations in the entity (consciousness and reality) are presented without clarifying or even questioning the guiding regard, that according to which they are distinguished, which is precisely being in its sense.[120]

Heidegger asks, with respect to the original asking of the question of being by the 'Greeks', if this neglect of being by 'phenomenology' is indicative of the 'history of our very Dasein'. He says that one can only express oneself, come into one's own in rebellion against this tendency of falling, in a struggle against the 'hegemony' of the idol of reason which Husserl takes from Descartes. This rebellion *removes itself* from the 'idea of absolute and rigorous scientificity'.[121] For Heidegger, not only is Philosophy not a *strong* science, it is not a science at all.

Radical phenomenology and the question of being

Philosophy, for Heidegger, is an originary tendency towards the 'matters themselves'. It must ceaselessly rebel against a deference to the falsifications of the 'tradition'. As a possibility, philosophy must be held open, not displaced, or fixed by a premature answer to the question of being. To hold open a tendency towards 'matters' is, Heidegger declares, a retrieval of an original posing of the question of being asked by the 'Greeks'. He assures us such a historical reference is not the repetition of 'authority', but a retrieval of a thread of questioning – understanding it by entering into it.

> The sole ground of possibility for the question of being as such is Dasein itself insofar as it is possible, in its discoveredness in possibilities.[122]

Heidegger is directing his listener towards the 'matters themselves' in an original apprehension of this being which either appropriates or discards that which has been, and thus gives rise to 'tradition' in the first place. The question of being is to be grounded in the questioning itself. In this way, the question of the 'being of the intentional' is surpassed by a more basic question of the meaning or sense of being, in the movement to matters themselves. Heidegger states:

> This question can be attained in any entity; it need not be intentionality. It does not even have to be an entity taken as a theme of science.[123]

For Heidegger, this turn towards the 'matters themselves' is a questioning 'to the very end or to inquire into the beginning . . . to allow entities to be seen as entities in their being.'[124] This circular hermeneutics of existence is, at the same time, a comportment amid the historicity of thought. Philosophy cannot be born 'mid-air'. In this way, the question of being is not simply uttered, but relies upon an initial arrival at the phenomena, which cannot be, as we have seen, an 'analytic description of intentionality in its *a priori*'. With his apprehension of the historicity of his own thinking, Heidegger inaugurates not only a radicalization of the 'thematic field', but also a 'more refined conception of the entity having the character of the intentional'. In this light, he indicates that it is *time* that will be our clue for the articulation of the question of being, since the 'history of the concept of time . . . is the history of the question of the being of entities . . . of the decline and distortion of this basic question . . .'[125]

The question of being 'must be articulated'.[126] Any 'answer' to such a question will point to something that is already 'there', understood in a question, an opaque, indefinite pre-understanding, but 'still an understanding'. This is the 'primal source' of the question, of expression, 'conceptuality'. The questioning begins in the indefinite, vague, in a 'cloud of unknowing'. This is not the question of Kant's deduction, of how to bridge the gulf betwixt sense and the concept, but of an indigenous expression that emerges amid the pre-understanding of everydayness where there is always an understanding of the 'is' but 'without being able to say more precisely what it actually means'.[127]

Heidegger lays out the point of departure for our questioning in a series of considerations which describes the character of finite knowing which emerges from the site of factical existence. The direction of the inquiry is guided by an 'inquisitive looking upon' the 'on-which' of the questioned in a cultivation of a 'preparatory outlook of the interrogating regard . . .' 'What is interrogated' can be anything of which we speak, or point to, take note of, intend, towards which we act. 'What is asked about' is a questioning of this being with respect to its own being. 'What is asked for' is a sense or meaning of being, expressed through a 'conceptuality pertinent to what is asked for . . .' Yet, all of these questions of access, direction, outlook, are, Heidegger tells us, all entities. There are myriad entities, phenomena. Yet, which being is to be questioned? What access do we have to this being? He sets out a provisional answer in the 'being of the questioning of the questioner himself'.[128] The questioner, in that he questions, is disclosed as the phenomenon. To the 'sterile' charge he moves in a circle, Heidegger responds:

> . . . the entity whose character is access, experience, etc. must be illuminated in its being, to the point where the danger of a circle exists. But this would be a circle of searching, of going and of being . . .[129]

The depth of penetration into this questioning being will project a horizon for the radicality of a response. The question is of a *sense of being* for an entity, this questioning of the questioner. In this way, there will be a detour via the questioning entity, a preparatory uncovering of the entity simply in 'what' it is, which Heidegger claims is 'the entity that we ourselves are', self-affected, temporal existence. He states, 'This affectedness of the questioning entity by what is asked for belongs to the ownmost sense of the question of being itself'.[130] Questioning is to be open to itself, a 'phenomenology of Dasein', an intimate openness to being. Radical phenomenology is a self-interpretation of existence of original temporality. Yet, there is a danger in self-expression, Heidegger warns us, that in our attempt to negotiate the peculiar 'doubling' of expression, that when one risks self-expression, there is a possibility of 'getting lost'.

Chapter 3

Temporal Expressions of Being-in-the-World

The fragile walls of your isolation, which comprised multiple stopping-points, obstacles of consciousness, will have served only to reflect for an instant the flash of those universes in the heart of which you never ceased to be lost.[131]

At the heart of the *History of the Concept of Time* ferments the temporal problematic of the *topos* of the questioner. In the second division of his lecture, Heidegger begins to set out provisional answers to the questions which he raised with respect to the coherence and radicality of phenomenology. Yet, he does not tell us the full story all at once. His indications in this lecture, and in the first division of *Being and Time*, remain *prospective* in the sense that he is laying out the results of his phenomenological activity as a way to point out the basic situation of Dasein. The indications point to that which is there and do so in a way which resolves the problems inherent in the early phenomenological movement. The whole story, including the emergence of this conceptuality from amid the horizons of ecstatic temporality, I will begin to tell in the next chapter where in a return to *Basic Problems*, I will trace the retrocursive origination of an indigenous understanding of being and self-expression from the root of original temporality.

In the following pages, I will explore one of the *early drafts* of the preparatory 'analytic' of Dasein, of being-in-the-world (**In-der-Welt-sein**), of factical, lived existence under the shadows cast by the unanswered question of the temporal 'origin' of indigenous 'conceptuality'. To put it differently, we will examine Heidegger's second answer to the question 'how being is understood and conceptually comprehended by means of time.'[132] Having indicated a provisional self-finding (**Befindlichkeit**) of that entity to be 'interrogated', *History of the Concept of Time* begins its phenomenology of existence amid being-here in the world, with one's 'familiar' absorption among beings, a 'world' of entities, things – and other 'selves', Daseins. The place of Anyone, in the familiar, is, moreover, organized in the manner of a 'by means of'/'in order to' so as to perpetuate and reproduce things and selves. There is thus a relation, as we have seen, between *logos* and being. We have a pre-understanding of this relation as the incessant 'talk' about beings within this discourse of *everydayness*, occurring simultaneously with an encounter with beings and non-beings, in a mode of uncovering. It will be from out of this vortex of everydayness that Heidegger will divine his existential indication of being-in-the-world and the existentiales which describe this *topos* of existence.

We begin with the insight that we already have an 'understanding' of our involvements. That misleading, 'sham question' of how a 'subject' comes in 'relation' to an 'object' or, of how a 'table' of concepts may be applied to an experiential manifold is blind to that which Heidegger has indicated as a bifurcated intentionality, the double relation of *logos* and being which I have just considered in Chapter 2. Epistemological questions, for Heidegger, are diversions from the original matters. Beginning in severance, they enact a theoretical falsification of the phenomena and suppress the expressive indication of 'things themselves'. For Heidegger, on the contrary, there already **is** a pre-understanding amid one's basic involvement, in-being, or, orientation.

Heidegger's *destruktion* of 'phenomenology' uncovered its tacit commitment to the meaning of being which is that of 'Reality', of natural science, of physics, of mathematical being. In this light, it was no surprise when these inconspicuous commitments gave rise, in the working-out of phenomenological philosophy, to unresolvable and 'vicious' problematics with respect to a sense or meaning and hence the limits of one's 'personal being'. Tragically, it was Husserl's unfinished phenomenology which allowed theoretical conceptualities and methodologies to suppress, conceal the phenomenon and the intimacy of its self-interpretation of finite existence. For Heidegger, a return to the root of phenomenology must take place as an existential analysis, a fundamental ontology, one that is 'prerequisite' to regional ontologies, or sciences, of this entity, such as anthropology, ethnology, political economy or theology.

Heidegger, in the *History*, which will find its way into the last chapters of the larger fragment of *Being and Time*, underlines original temporality as the sense, or meaning, of these characters of being, of the existentials, the indicative structures of being-in-the-world. He is very clear that a 'hermeneutic phenomenology' must have a 'fundamental ontology' as an appropriative situation of 'grounding' that reveals the temporal singularity of thought which is founded upon 'no-thing'. Phenomenology cannot be removed from basic phenomena and horizons for phenomena without losing its orientation and its sense of cohesiveness for its project as a whole. That such a basic projection of being upon temporality, or in other words, such an understanding of being, is 'intended' even in the analytic of being-in-the-world, of Care, can be seen in *Being and Time* and in all of the lectures of the period, such as in *Basic Problems*, where the characters of being-in-the-world are disclosed as projections of an original, ecstatic temporality (cf. Chapter 4). Such a radically temporal interpretation gives back to Heidegger his 'love and hate'.[133] With such a retrieval, thought can again exhibit the ecstatic movements of a disclosive mood, a disposition that discloses original phenomena, as in his indication of anxiety.

The orientation of being-in-the-world (In-der-Welt-sein)

In his discussion of the spatiality of the world in the *History of the Concept of Time*, Heidegger contests Kant's essay, 'What is Orientation in Thinking?' in its contention that an orientation is fathomed by a 'feeling of a distinction between my two sides'.[134] This 'feeling', reminiscent of Rickert's assertion that there is a 'feeling' of the truth of a judgement, signifies an *a priori* grasp of orientation, seated in an autonomous subject. As with the case of Rickert and his 'mythology of consciousness', Heidegger denies the phenomenological adequacy of this account of orientation, since a subjective 'feeling' and expression of this 'feeling', are already dependent upon a *being-in-the-world*, from which orientation 'first' becomes specified amid a temporal, worldly existence.[135] Heidegger contends that Kant went radically astray 'due' to an 'inadequate concept of dasein', of existence and of the world. As we will see in Chapter 5, Kant's conception of a rationalist subject which 'posits being' is a laceration of the self from its world, from original temporality, a severing reification that seeks only to prop up a questionable 'authority of reason'.

Heidegger begins his radical phenomenology of original temporality with the self amidst its being-in-the-world, with the finite existence of the questioner, and of its self-interpretation amid his or her 'there', as an orientation, *sich auf etwas verstehen*. He suspends the uprooted solutions of the theoretical 'tradition', so as to apprehend the 'matters' to which the tradition is inherently 'blind'. The question of the authority of reason, as Beiser clearly shows, asks after the legitimacy and power of reason in its theoretical and practical authority over the characters of sensibility, imagination and temporality. Yet, Kant innovates with respect to the grammar of reason with his 'severance' of the theoretical from practical reason, which is seen to be a solution to the embarrassment of 'Spinozism'. This innovation moreover is executed without undermining the authority of reason. The territory of practical, moral freedom may be posited beyond the proper limits of theoretical reason, in order to 'make room for faith'. Yet, despite its mere practical status, reason in this sphere has the power to directly determine the Will. 'Practical' concerns a 'kingdom of ends', noumenal, unconditioned, no-thing with respect to 'knowledge', one which is orchestrated as a practice of morality. Set free from the requirement of a cognitive certainty, practical reason perhaps has more worldly power than the theoretical face of reason. Despite the stated differences betwixt these faces of Janus, each is a definite aspect of reason deployed in differing spheres of 'relevance'. In this way, a 'free-floating' feeling for left and right which Kant expresses, without any concern for an orientation amid **world**, is indicative of his basic 'allegience' to an autonomous reason. Kant defers the question, postponing the possibility that the faces of reason are in truth 'rooted' in transcendental imagination, or, more adequately indicated, Temporality.

The reference to Kant's essay therefore is not a retrospective taking sides with the eighteenth-century 'imaginists', such as Jacobi and Hamann, or indeed Hume, but is instead a statement concerning the basic meaning of thought, of questioning the 'conditions of possibility' of finite knowing. In his interpretation of intentionality, Heidegger thinks along with Kant (and Husserl), to the extent that 'experience' is invested, as we saw in Chapter 2, with categorial structures. Yet, what is in question is not a relative position of faculties but the 'origin' and 'operation' of these categorials. As Heidegger has suggested in his 1928 logic lectures, 'rules' for interpretation need not be 'grounded' upon logic, nor can they be, in that 'logic' is solitary. Any reason, or understanding, divested from sensibility, and therefore, the world, temporality, imagination, is not and cannot be a 'first'. For Heidegger, being-in-the-world is the 'first', a basic 'in-being' of involvements, and perhaps, of interests. As a basic predicament of existence, indeed, 'in being' is an *a priori* situation for finite thought. Admittedly, there will emerge in this circumspective understanding a degree of separation betwixt this knower and that which is known. Such a separation, however, though always in danger of getting lost, would exhibit the unity of lived intentionality within a context of orientation guided by this 'unity' of the phenomena. Heidegger, echoing his work on Duns Scotus, seeks to enter into the situation of the 'event', **this**, 'haeccitas' – not to 'know' the 'event' but to understanding by being the event. *This* is the 'ground' – 'time is its own norm'.

Being-in-the-world, tracing its genealogy from Dilthey's 'holistic *a priori*', is a *topos*, place, of existence, not as a 'model', 'norm' or 'theory', but a context, a complex indication, pointing to that which is 'there', not objectively, but phenomenologically. Contrary to Husserl's brand of philosophical egoism, Heidegger deems it impossible for there to not be a be-ing *in the world*. There can be no 'ipse-ity',[136] of regard for the self, without 'communion', or, be-ing in the world. This indicates, as Sallis suggests, the importance of the transcendental imagination with respect to the 'basic problems of phenomenology'. It is this power (**Kraft**), which as a surrogate for temporality in Transcendental Philosophy, projects one's own temporal world in its ecstatic 'unity'. Paradoxically, it is with uncanniness that the world is not only susceptible to an utter derangement of the ordinary,[137] but can also trace its event of projection.

A hermeneutics of temporal existence

Heidegger reminds us that, for the most part, we are absorbed amid the everydayness of 'generic time', the concerned and understanding involvement 'within' our world, and the 'duration' and 'space' of its operational functionality. We are all mere 'appendages to the machine', and this is widely held to be 'normal'. We 'fall' towards this normality, and 'consent is manufactured'. Amidst one's absorptive involvements in this 'hegemonic' and 'legal' workworld of continuous flow production, of 'commodities by means of commodities',[138] one finds oneself amid its involved proximity, one encounters one's world. To a discourse of everydayness, as a language which ceaselessly innovates and repeats itself so as to hold us captive to an image, to an indicative nexus of 'normality', there is related a controlled environment, amid which the self is lost in the Average, as it falls into idle talk, into curiosity, ambiguity. In its repetitive familiarity, everydayness discloses what is there *for itself.* What it does not point to is covered over by incessant indications, via assimilating, disorienting and distracting jargon. At the same time, there is also *something else*, beyond the concealment of the 'fact', there is also the eruption of the overwhelming, of a no-thingness which ex-ceeds 'normal' limits of everyday familiarity. As a threat, it is incomprehensible, welling up beneath this site of the ordinary as the uncanny.

We begin with that which is nearest, but, we attempt to come 'near' to that which 'normally' is farthest. The 'first approximation' is an attempt to lay out, in a provisional way, the 'basic constitution' of the questioner, spoken as the term Dasein, as being-there. Working against our indigenous tendency to falsify existence, Heidegger retrieves an original sense of truth and falsity, as a play of unconcealment and concealment, and which displaces the hegemony of the propositional sense of truth as judgement. For Heidegger, truth is obvious, we are in the truth, falsity is concealment and concealing. Both dwell amid a basic situation of existence. Indeed, with Nietzsche, there are errors, we live errors, in errancy. Yet, amid this normal errancy there are moments of vision, truth events, radical breaks amid system, eruptions: revolution, poetry, art and *events of questioning*, which for an instant disclose the truth of that which is 'there', or indicates that another truth may be possible.

However, and this is radically endemic amid the modulations of temporality and historicity, as soon as an event breaks open it becomes falsified, lost in the discourse of everydayness, of the Anyone, and of other events. Yet, there is 'no one else to blame', as this is a destinal flight of Dasein from itself, a 'cover up' of its ownmost truth. There persists amid this existence, a 'restless exchange'[139] between Anyone and this intimate sense of oneself. Yet, that which is significant in this exchange is not to set up another 'dichotomy' between an Anyone and oneself, but to see this exchange as the way of being of this temporally singular, indigent existence. 'Truth' as *a-lethea* is that which shows itself from itself, or, it is the phenomenon which erupts or is uncovered, amid the horizons of that which *must* remain concealed. To speak about this truth, since Dasein is 'in the truth', is to articulate a 'discourse of truth'. We are 'in' this world, involved amid our world, there with others in a world where there is no strict, logical identity, none of us can be so neatly segregated from the others. I cannot cut myself off from others, we are here together, even if I were alone as a hermit in the woods, I am still being-with (**Mitsein**): there is no escape from this other, nor need this be a problem. Yet, a more fundamental question emerges: how do I **know** this, why did I not simply remain assimilated, absorbed in the belonging of the familiar, 'being involved with' the Anyone? Why has questioning become accentuated at all, as it is not at all obvious why there would be any 'philosophy', any questioning if there were simply absorption, inconspicuous familiarity, dwelling as habitation, habituation, forgetting? What does forgetting entail? What has been forgotten? Or, in other words, 'what' is being remembered in the questioning? If that which is, is an indistinguishable play of truth and falsity interpreted in the sense of *a-lethea*, can there be a fixed standard, a measure, if all is embroiled in temporality, the self -projection of indigenous existence? The problem is that for us to ask a question of being-in-the-world, there must have already been a pre-understanding, which became 'disrupted'. In Heidegger's sense, in a twist of Leibniz's maxim that every statement is an 'eternal truth', existence is on 'the outside', in the truth amid the self-showing of phenomena. Heidegger does not simply presuppose the possibility of his own questioning. He points out antecedents in the tradition, but more originally, **he questions**, enters into a questioning, *before* he can provide an account or justification of its *de juris*. This is the case moreover in that this process must be 'performative' as it concerns a questioning in situation. This was the case in the events of his lectures.

Finite understanding has emerged as a temporalization, amid historicity: there must have been the occasion for an accentuation of this questioning itself, of the questioner, *if that is*, this is not to be another 'theory', *if that is*, we are to 'know ourselves' as an 'event' amid the world. Heidegger, in this light, sets out an initial hermeutical situation as that of the meaningful nexus of the workworld. It is a situation of 'meaningfulness' (**Bedeutsamkeit**), in the sense of an operational 'know how' (*techne*). This indication of a workworld is a provisional sketch of the non-cognitive context of the questioner. In this way, one understands what one 'knows' amidst that context in which one lives, just as in Wittgenstein's indication of 'language games' and 'forms of life' in his *Philosophical Investigations*. This world of work occurs amid an equipmental contexture where that to be performed is there *ready to hand* (**Zuhanden**), as for instance, a spoon or a book. I do not think, 'This is a spoon', but pick it up and stir the soup before it burns. In other words, amidst the lived situation of be-ing, a thing as a thing exists *per* flow of action, in the event. A temporal situation moves towards its 'fullfillment', in whatever crooked path this may occur.

All that I begin with is my mood, my disposition and my pre-understanding amidst and as a disclosure of my world. I begin writing letters, words I type upon this keyboard in a rhythm 'by means of' a language expressing 'my' thoughts as they 'come' or are 'chanced upon' – now I have begun to write about writing in a 'self-conscious' manner, my writing, even this that you read *now* has become thematized, it is under scrutiny. Perhaps, we should take a break . . . This is an example of departure from the absorption of reading and writing, equipmental

activity and this rhythm, of an 'event' of writing and reading – to a pause. This
pause intimates an instance of the emergence of the *vorhanden*, of a theoretical
perspective. Such a thematization becomes possible in the wake of an interrup-
tion of concrete activity, a breach in the familiarity of involvement, of typing
and reading as action – I stop – I sit, try to think of a relevant analogy, for example, I attempt to get back into it, to leave the theoretical perspective behind and
simply write. Suddenly, an intention is fulfilled, an example 'erupts', is there as
if from 'nowhere'. It is a familiar example, amidst its absorbed involvement, it
becomes inconspicuous.

Heidegger's example is a hammer. One does/can not notice the hammer
per se, one grabs it in the moment and strikes, one holds the hammer, my hand
becomes numb amid this repetitive motion. Although we are allegedly 'awake',
as we are reading these words, each of us can be lead to remember this involvement of understanding. Indeed – one types upon a typewriter, prepares food,
cares for a sick child in the night, buries the dead. An 'Anyone' works amid
'equipmental' contexts, situations of meaning, temporal 'worlds' of 'historical'
action. Yet, how can I 'know' this? How have I encountered world as world? We
have already given a clue to an answer: one only knows 'due' to one's own
breach. Heidegger contends that we can only know our own self when it has
been resisted, broken or has encountered a limit-situation, via which each finds
herself in her 'truth'. Normality suspends . . . with an eclipse of the sun, an
earthquake, a flood, the death of another – a truth event. Amid our existence
there is an 'uncultivated' 'raw' understanding gathered as a 'makeshift', a provisional shelter for 'rules of thumb'. Finite knowing is a 'founded' mode, its
possibility is exposed in a breach, upon an abyss, a groundless act, it rests upon
that precipice of one's ownmost temporality. Such a situation of understanding
is rare, but is constitutive of our existence as temporal beings. Such a 'moment
of vision' in which one glimpses one's world is not, for Heidegger, to be conceived as nakedness-before an absolute, nor as a glimpse of 'objectivity', if only
for an instant. Neither is it 'subjective'. As we have already seen, he has already
suspended 'positions' that sever themselves from the original opaqueness of
being-in-the-world. It is in the midst of these modalities of falling, idle talk,
curiosity, ambiguity, that Heidegger begins to indicate a pre-understanding of
finite existence. In our initial awareness, there is a dispersed focus amid this
chaos of temporal happenings. There is not simply a single 'truth event', or not
usually. Instead, there is an eruption of events, ceaselessly. It is not as if we can
isolate synchronic structure from diachronic operation – at each stage of
Heidegger's analysis, we are reminded of this impossibility.

Heidegger gives us no redemption, no escape from the vortex of temporal
fluctuation, from the radical temporalization of mood/thought. He conjures
forth a vision of existence not dependent upon the 'myths of consciousness'. As
he expresses in his early and later writings and lectures, truth is always amid

untruth in the play of a-lethea, a play from which emerges a 'clearing'. But, this is not a clearing which, once and for all, casts out the opaque refusal, but one that 'holds' itself in the 'incomprehensible'. Amid this *makeshift*, there is no autonomous subject, but we may be in some sense *sovereign*, in the sense in which Bataille has intimated. At the end of the day, we are not in control, nor, do we posit 'being'. As radically temporal, we have only an uncertain grasp of that which is the case. With Socrates, we must confess ignorance. Yet, it will be on the basis of this confession that we will be open to the truth.

Chapter 4

Ecstatic Temporality and the Meaning of Being

An affair that is cleared up, stops, it is something closed to us. What did the god mean who advised 'You know yourself!' Perhaps, he meant, 'Stop! You are something closed to yourselves! Become objective! – And Socrates – and the "scientific man"?'[140]

An unexpected breach shatters our picture of substantial unity, and, as with Hegel and Nietzsche, Heidegger enters into the question of the power of 'negativity', or the 'nothing' amid our understanding and sense of being. Indeed, he raises the question of the 'that' and 'how' of that breach, which throws us out of inconspicuous familiarity. This question will become, in *Being and Time* and in the later lectures, that of anticipatory resoluteness, which is revealed as an abyssal 'ground' of transcendence itself and of its self-interpretation and expression. With this tracing of expression and interpretation to their source in a temporal event, the prospective indication of the characters of being in the world, which I set forth in Chapter 3, will be considered in this chapter from the retrocursive perspective of their origination in ecstatic temporality. It is this 'return to the source' that we will explore in the following pages.[141]

Picking up again our investigation of *Basic Problems*, I will lay out Heidegger's tracing of the conditions of emergence for an understanding-of-being in original, ecstatic temporality. In other words, we will begin to answer our question at the close of the first chapter concerning an 'origin' of understanding in a projection of ecstatic-horizonal temporality. An 'answer' to this question can be found, Heidegger claims, in un-covering the ways of being, characters of being-in-the-world, as 'thrown disseminations' of the projections of ecstatic temporality. The un-covering is catalyzed by the event, a breach, disruption, radical absence. It is a *going beyond* one's familiar absorption in this world, an event which is the source for an original disclosure of the characters of existence, of being-in-the-world. In the anxiety of one's own being-towards-death, one's familiarity shatters in the wake of the uncanny, the overwhelming.

Understanding and temporality

In the *Basic Problems*, Heidegger contends that understanding, contrary to the 'mythology of consciousness', is a 'comportment toward beings'. We encounter beings, however, 'only in the light of the understanding of being'.[142] An understanding-of-being (**Seinverstandnis**) is 'there' with each and every specific comportment, whether it is of the theoretical-cognitive (**vorhanden**) or practical-technical (**zuhanden**). In this way, understanding is an 'original determination of Dasein's existence'.[143] This is not the theoretical understanding of Kant, immune from time and existence. As being-in-the-world, as 'there', Dasein, while exhibiting a specific understanding-of-being, is, at the same time, occupied with its own 'ability to be'. Da-sein is concerned with its ownmost possibility: not empty logical possibility, but its *eigentlich* possibility, which is constrained by temporal horizons and expressed as its own characters of existence. Dasein is a being free for a basic understanding-of-being. Heidegger states:

> To be one's own most peculiar ability to be, to take it over and keep oneself in the possibility, to understand oneself in one's factual freedom, that is, to understand oneself in the being of one's ownmost peculiar ability-to-be, is the original existential concept of understanding.[144]

Understanding, as we have seen, has been initially discerned in the midst of a concrete 'operation', as with a skill or a cultivating. Yet, with the breach, the understanding deepens amid a primal openness to one's own possibility. This possibility of the self exists only in the projecting of oneself upon this free possibility, to keep, hold oneself, within this possibility, and thus, to unveil oneself as this possibility. This temporal possibility is an event of concrete self-understanding, an 'authentic meaning of action' amidst one's own 'historicality'.[145] Understanding, in a displacement of the Kantian understanding, *vorstellen*, an isolated faculty of rules, is, for Heidegger, a 'basic determination of existing'.[146] With understanding, one's own existence 'temporalizes itself', one projects an understanding upon one's possibilities. Existence as this understanding, as involvement, is that which has been 'chosen' from a finite situation, which is a myriadity of possibility. Each projection, as an understanding-of-being, unveils itself, in each instance, as a possible being-in-the-world, an ability to be in a world, and equiprimordially a being with others and a being among 'intraworldly beings'. Ecstatic understanding is a free act and 'has the intrinsic possibility of shifting in various directions'.[147] One could shift towards a breach in which the 'There' of existence is disclosed, or one could shift to the succumbing of oneself to rhythmic plays of things, to cover up one's ownmost possibility. One is also free to be inauthentic, to be 'in the untruth'.[148]

An understanding-of-being haunts each act of disclosure of the being of beings, and that of others and of things amid this world. Heidegger, in character, states that 'at first', this understanding-of-being is indifferent with respect to the specific character or sense of being of the being in question. Yet, in this indifferent, pre-theoretical and pre-practical pre-understanding of being-in-the-world, there is projected not only the being of beings but also 'in some way being as such'.[149] In the very act of questioning, there is disclosed the be-ing of the questioner, and 'in that process being is understood'.[150] As pro-jecting, an understanding-of-being must have its own *upon-which*, it own meaning that is 'at first' obscure. Or, we understand a being by projecting it upon its being. However, this latter, this being, if it is to be understood, must also 'have' its own upon-which of projection. In this light, Heidegger sketches a provisional answer to the question of an 'origin' of understanding:

> If Dasein harbors the understanding of being within itself, and if temporality makes possible Dasein in its ontological constitution, then *temporality* must also be the *condition of the possibility of the understanding of being* and hence of the *projection of being upon time*.[151]

Temporality is the 'condition of possibility' for an understanding-of-being, an 'original determination' of the existence of Dasein. The understanding-of-being of Dasein, Heidegger continues, occurs through the projection of being upon time. Yet, as we will see, being is projected upon time, not as the regimentation of existence through a matrix of unworlded characters of being, regarded as concepts, but instead in the sense that these characters, as expression, are themselves revealed as projections of ecstatic temporality. This understanding is to be understood in light of the temporal horizon *upon which* one's understanding-of-being is projected.[152] That aspect which is 'beyond' being is that upon which being is to be projected. Understanding, or the projecting-upon, envisages its horizonal schema as that upon which it is projected, the background of its sense or meaning. The *a priori* light of an understanding-of-being is the intimacy of self-interpretation, of questioning – it illuminates that which is to be encountered. For Heidegger, if we are to understand, it is into no-thing, finitude that we must look. He declares in a quite mystical tone:

> The basic condition for the possibility of understanding the actual **as** actual is to look into the sun, so that the eye of knowledge should become sunlike.[153]

The Sun is the good, which is 'hardly to be seen'. It is the 'good' which determines this understanding-of-being, for the 'Greeks', a temporal sense of being, a 'for the sake of' *eneka* (ενεκα) that 'guides' an 'in-order-to' *este* (εστε). Finite receptivity and openness to that which guides our temporal being is that which overwhelms existence and the place to which one is thrown after being extinguished by a nearness to the light. Finite understanding is thrown from this gaze into the Sun. It comes *ecpyrosis*, out of the fire.

The question has been expressed by Heidegger in other ways: How is understanding 'rooted' in original temporality as its situation of possibility? How is it rooted in original temporality, which is the 'ecstastic-horizonal unity of future, past, and present?'[154] If this Sun is, in Heidegger's case, the *a priori* of temporality, a projection of being upon a horizon of temporality, an understanding-of-being in terms of time, how is such a makeshift understanding possible? We have already laid out the point of departure for an answer in the last chapter as an understanding-of-being within the horizon of everydayness. Such an understanding-of-being occurs in a projection of oneself, of one's understanding upon one's everyday pre-occupations. In this way, an understanding-of-being is disclosed as a comportment of everydayness, of 'Anyone'. Heidegger states, 'The dasein thus comes toward itself from out of the things.'[155]

In such a situation, the freedom of Dasein, in Heidegger's sense of a self-temporalization, is held captive to a proliferation of things amid the everyday leeway of encounter. In this way, we are dependent upon 'things' and upon *what* 'things' will bring to us. We are merely expectant, which Heidegger calls the temporality of an inauthentic future. He claims that we have forgotten our original sense of being, that which is true about ourselves has been covered over in that we, in a similar way as the dependence of early phenomenology upon a natural scientific worldview, have no means of differentiating ourselves from mere things: we have been swallowed by a homogenous monism of entity.

However, in a reversal, he states that forgetting, as annihilation, is a derivative mode of forgetting. Speaking to the possibility of retrieval of an *eigentlich* understanding, Heidegger states that there is an original forgetfulness that is a mode of ecstatic temporality. Original forgotteness is a having-been-ness which remains as a mode of the being of Dasein as existing, in a manner similar to Brentano's spectral 'primordial associations'. The forgetfulness is not a loss of that which has been as in the derivative mode of forgetting. Instead, for Heidegger, forgetting forgets itself, it becomes open to its forgetting of itself, not as annihilation, but as remembrance which holds open a horizon of futurity, embracing this movement of disclosure, of these uncovered possibilities of existence. Pointing towards his earlier analysis in *Being and Time*, which is 'presupposed' in *Basic Problems*, he announces that it is 'resoluteness' which is

> . . . our name for authentic existence, the existence of the Dasein in which the Dasein is itself in and from its own most peculiar possibility, a possibility that has been seized upon and chosen by the Dasein itself. Resoluteness has its own peculiar temporality.[156]

This is not 'where' we begin, however, but instead, we find ourselves amid a nexus of 'handy' things and of a concomitant understanding-of-being which is contained in the very design of these things. As these things enter into our world, we encounter them in the light of our existential understanding, and thus, our commerce with them is 'grounded in a specific temporality of being in the world'.[157] In order to deploy, use this equipment, gear, we must project the tool, the handy thing, upon a 'functionality relation'. We already possess an 'antecedent understanding of functionality', of a letting function. This understanding expressed the 'with which' of equipment by which a 'for which' will be fulfilled. A letting function is a temporal understanding, a play of projections of retention and expectation amid a site of en-presenting, not as a 'now', but as a clearing, not the 'here' as an uncomplicated, *naive* present, but a *topos*, thrown from an ecstatic temporality. One's self exists betwixt this play of ecstases of retention and expectance, amid a fleeting present in between differing senses of being, and its rules of 'in order to', 'for-which' and 'with-which'. Temporalization comes from out of the 'middle'.

Through this 'middle', understanding is projected, in Heidegger's metaphor, as an open horizon amid the light of which being will be encountered. Yet, this thingly light and this understanding-of-being and temporality, as it is 'grounded' on the situation of the handy, of gear, is not yet an uncovering of an original temporality. The equipmental contexture is an opening of a meaningful horizon for an entity, a pre-understanding and point of departure for a self-interpretation of existence. In this provisional phase of phenomenology, any question of the being of differing beings is left 'indifferent, unarticulated'.[158] At the same time, however, Dasein also has an understanding of the 'for-the-sake-of' in which its own self is involved and concerned about its own possibility. It feels the gnawing question of the sense of its own being. As I have suggested, Heidegger indicated an ontological determination of Dasein as that which is concerned with its own ability to be. He is concerned to disclose this ability to be of the self as the 'first', as a temporal *a priori*, as the condition for the uncoverance of a 'world' of our communal and singular be-ing. From this perspective, one's thrown projection of world is the original possibility for a communal disclosure of 'world', and thus, of communication, expression. In this way, the world is not an aggregate of world-views, of ideologies, or, of an inter-subjective consensus, but, is an original 'unity' where any singularization finds its source of be-ing and its place of expression. Heidegger states:

> Self and world belong together in the single entity, the Dasein. Self and world are not two beings, like subject and object, or like I and thou, but self and world are the basic determination of the Dasein itself in the unity of the structures of being in the world.[159]

It is this selfhood as a being concerned with its own ability to be that is the 'ontological presupposition' for selflessness with respect to an I-thou relationship. Heidegger does not withdraw however into some windowless turret to shelter himself from the storm of the world. As being-in-the-world, Dasein has the character of be-ing which is finite transcendence, not in the sense of a substantial 'God' or of that which lies beyond the subject as an object which the former can *choose* not to choose. On the contrary, Dasein is overstepping as such, transcending, be-ing *already always* amid its 'world'. Such a being discloses the condition of possibility for a pre-understanding amidst the factical world. Understanding, as the light of world, is the *a priori* of any specific worldly encounter. As the overstepping, as transcending, Dasein, as the questioner, is 'not the immanent', as with the 'mythology of consciousness'. Heidegger states, 'The Dasein is its Da, its here-there, in which it is here for itself and in which others are there with it . . .'[160] It is only in light of this characterization of Da-sein as transcending, that we can begin to answer the question of the 'indifferent, unarticulated . . .' in which all beings are determined according to an understanding-of-being that is derived from an objectification of worldly things. Only Dasein is transcending, its being is *different* and *articulated* – this is its *ontological difference* from a thing. Dasein is 'beyond itself' in its being-in-the-world, and thus, it understands itself either from 'things' or from its own 'nothing'.

With this indication of self-understanding of the *transcending self*, the question of the meaning, sense of being, refers back to this be-ing of Dasein as being-in-the-world. Existence is on the 'outside', 'amidst', it transcends towards itself, others, and to 'things'. This transcending of Da-sein possesses an 'original familiarity' with 'world'. But, this is a familiarity that disguises its radical non-knowing amid the thrownness of ecstatic temporality. As the ecstases of temporality are projections of finite transcending, each questioner projects, at the same time, its own horizon, as a removal, 'whereto' that opens, releases a horizon for a rough sketch, a projection of a being upon its be-ing. Being-in-the-world as transcending is 'founded' upon an 'original ecstatic-horizonal unity of temporality', and since understanding-of-being is possible only amid one's finite transcending, the 'answer' to our original question is that the understanding-of-being is rooted in temporality as its situation of possibility.

The projections of ecstatic temporality

In the previous section, we have laid out Heidegger's provisional answer to the question of the relation of understanding and temporality in the indication of the latter as the situation of possibility for an understanding-of-being. We have also examined Heidegger's disclosure of the unity of thought and being in being-in-the-world, having dispelled the 'mythology of consciousness' with its domesticated and homogenized time. Linear time, for Heidegger, is blind to factical, lived existence, to the specificity of being-in-the-world. It is a second order fabrication, an 'unworlding', and in order to make this clear, he seeks to disclose explicitly the existential rooting of the previous understanding-of-being of the handy, *zu handen*, in original, ecstatic temporality. Temporality makes possible an understanding-of-being, but without an everyday Dasein 'knowing about it explicitly'.[161] In this light, the being of the ready-to-hand, as it is always involved in an understanding-of-being, is already projected upon temporality. This does not mean that a 'relation'of being and time is hidden. Indeed, such an intimacy is explicit in everyday life and in the historicity of its articulation. It is only through an immersive understanding in the act that we can truly understand the specific 'working' of equipment, and this means that only if we are 'there' can we 'reproduce in thought the factical commerce with these things'.[162] Each 'understands' very little – we understand 'what' is familiar, proximal to our lived existence. For the most part, however, the 'world' consists of the unfamiliar, having the character of a 'privative mode of uncoveredness' comprehended from the perspective of the familiar.

Heidegger begins not with the artificial scenario of the theory of knowledge, but in our familiar immersion in available, everyday involvements, of a letting be, or amid a be-ing thrown into a breach, or the unavailable. Availability and unavailability, presence and absence, are, Heidegger suggests, 'modes of being' of a 'single basic phenomenon' of 'Praesens'[163], which is an indication of a meaning of being which points to the rooting of these ways of being in original temporality. *Praesens* points to a temporal ground for an understanding-of-being as such, a *topos* that is the 'unity' of the three ecstases.[164] It comes from the 'outside', and is to be the key to an understanding of the handy as intra-temporal. *Praesens*, in a way which has not been specifically clarified by Heidegger in any of his published or unpublished texts, presupposes a specific 'event' of awareness which exceeds generic time and understanding. Yet, as there needs to be a preliminary consideration of the projections of the ecstasies as such, I will postpone further consideration of *Praesens* to Chapter 9, 'Transcendental Imagination and Ecstatic Temporality'.

An ecstasis is a removal – not to 'nowhere' – but a projection upon a horizon, which in this way fulfils its temporal be-ing. Praesens, designating in this instance the horizon for the ecstasis of the present, is an *upon-which* for a projection of one's be-ing upon time. The sense of the present is 'beyond itself', in a projection which 'has within itself *a schematic pre-designation* of *the where out there* this "beyond itself" is'.[165] In this way, the ecstasy of the present understands itself in a projection upon its own horizon of *praesens*. Amidst this *projection upon*, the present is open for an encounter with entities, which are themselves understood '. . . antecedently upon praesens'.[166] It is in this way that we can explicitly see a link between an understanding-of-being and temporality, between entities whether present or absent, as modalities of presence, and the horizon for this ecstasis of the present, *praesens*. In this light, Heidegger 'answers' his own question of a 'relation' of being and time: '. . . we understand being from the original horizonal schema of the ecstases of temporality.'[167] One must not however imagine that the horizonal schema are 'out there' and the projections reach out to them as if they were something distinct. These schemata, Heidegger declares, are not 'detachable'. Temporality projects these schema, the 'unity' of which allows for the 'clearing', the 'there' of openness, and thus, for and understanding-of-being. The horizonal schema are 'united' as are the ecstases, but its contours of figuration shift via one's prevailing temporalizations. In this way, 'looking backward', retrocursively, the understanding is possible in a self-projection of temporality. Heidegger states:

> Temporality exists – ist da – as unveiled, because it makes possible the 'Da' and its 'unveiledness in general.'[168]

As ecstatic self-projection, temporality unveils *praesens* in a non-conceptual manner and is the 'ground' for transcendence and for understanding. But, this means that 'time breaks through', that in projecting as understanding, there must be some understanding of temporality, even a mis-understanding, as in the case of a 'generic time'. Temporality is 'there' in our 'here-there' of commerce with things, even if there is no thematic apprehension of this 'presence'.

In a look back, Heidegger states that each of the projections, not only those of the ecstases, but also that of being itself upon time, have an 'end' in an ecstatic 'unity' of temporality. He states, 'But this end is nothing but the beginning and starting point for the possibility of all projecting.'[169] Temporality is the beginning and the ending of Being – it *is* the possible. It is higher than anything that may arise from it, of anything actual (falling). Heidegger states that all origination in the ontological is a running away from the 'superior force of the source',[170] pleading to be allowed to return to one's own finite patch of existence. Each must understand itself, in a way similar to Plotinus, from the 'perspective' of that which is closer to the source – yet, in its 'deficient similarity'.[171]

Heidegger states that we can often understand something from the perspective of its privation. Since temporality has 'negation rooted in it', a temporal understanding of the being of handiness will find its possibility in an 'orientation toward non-handiness'. Again, this mode of the handy has the character of an equipmental familiarity, of an unobtrusiveness. Or, in other words, amid the nexus of our concerned and understood involvements, we, for all extents and purposes, 'forget' what is *there* in any real, thematic sense. Amidst uninterrupted familiarity, we have an accessibility and understanding of this situation that has a 'peculiar' temporal character which allows us to lose ourselves in this situation, to become absorbed in it. On the contrary, it is that temporality of breakage, disturbance, of the obtrusive in the midst of one's own absorption that gives an 'orientation to non-handiness', to this no-thing.

Heidegger gives, as an example, missing that which one presumes is always already there. It is unavailable, which for him, is a mode of enpresenting. It is *there* since we expected it to be *here* – we need it, and our current equipmental involvement has been disrupted. In fact, our involvement has become this-missing-it (a past that bleeds), and not merely a theoretical attitude of a dis-function. As a disquieting 'presence' in that which is 'there', the absent *means something* amidst one's factical understanding of 'letting function'.

Heidegger states that the *praesens* of the missing has a horizonal schema of *absens*, as a modification of itself. There is a modification of a horizonal schema amid a self-projection of original, ecstatic horizonal temporality, which sets forth the 'conditions of possibility' for a finite understanding of be-ing, and that means a comportment with the being of the handy. It is upon this interpretation of the *a priori* that intentionality itself is possible. Such a *breach* throws the being of the handy into an unfamiliar conspicuousness, and it is in this way that a 'nothing' is included not only in temporality, but also, from what we now know, in the constitution of understanding itself, and of its temporal 'meaning'. It is the inevitability of rupture that projects the possibility of meaning, as it is projected upon this horizon of one's own future. It is in the rupture that one can turn from a self-understanding contrained by things, and instead, to recognize one's own ontological difference as a finite, transcending self.

Taminiaux questions the idea of a 'fundamental ontology' in his work, *Heidegger and the Project of Fundamental Ontology,* in his claim that Heidegger's fundamental ontology 'privileges reduction to deconstruction . . .'[172] He writes: 'it is from the Self that the call emanates, it calls Dasein to face its own Self, it summons it silently but exclusively to its ownmost potentiality-for-Being, and the call presents Dasein with nothing else for it to hear.' This criticism, recently repeated by Weatherston,[173] contends that fundamental ontology obviates the dimension of receptivity, of the givenness of the given amid the *mineness* of fundamental ontology. As I will seek to show, however, Heidegger's 'reduction' exhibits an intrinsic dimension of receptivity, one that *prioritizes* a radical openness to the phenomenon of existence. He has sought to excavate theoretical and practical conceptualizations in their cover-up and suppression of an understanding-of-being, of the radical finitude of existence. Indeed, in his criticism of the spontaneist reason of Kant, Heidegger seeks to unearth temporal being-in-the-world as the horizon for the constitution of the self. In this sense, his 'reduction' to his own original, ecstatic temporality is the emergence of a dimension of receptivity which is an openness for the necessity of 'otherness', for *Sein, Mitsein* and *Mitdasein.* It is the *es gibt* of a 'horizon' for a thrown projection of being-in-the-world.

In that it is thrown and finite, Dasein is receptive and open to that which enters or is thrown into its world. Yet, amid the historicity of a self, there are instances of disclosure which overwhelm existence, that withdraw from us access to 'normality'. Such instances are 'basic' as they are concrete encounters with the finite possibility and meaning of being. It is through such a rupture that a self finds orientation amidst his or her inexplicable and perplexing situation, being alive temporarily amid the absurd possibility of one's own impossibility. That about which Taminiaux keeps silent is the fact that each of us must die *my* own death, that 'I' must ultimately, though provisionally project myself upon *my* ownmost possibility. A 'reduction' is not a mere repetition of the Cartesian ego or a Kantian subject, but a disclosure of being amid the horizon of *my* world, a finite *topos* that shelters as a dwelling. In this way, as Heidegger states in *Basic Problems*, a reduction is a setting-free of an original place for finite thought. It is the task of this finite thought to clarify the specificity and meaning of one's own existence.

Chapter 5

Kant's Thesis about Being and Existence

This enclosed, segmented space, observed at every point, in which the individuals are inserted in a fixed place, in which the slightest movements are supervised, in which all events are recorded, in which an uninterrupted work of writing links the centre and periphery, in which power is exercised without division, according to a continuous hierarchial figure, in which each individual is constantly located, examined and distributed among living beings, the sick and dead – all this constitutes a compact model of the disciplinary mechanism.[171]

For most of his life as a thinker, Heidegger was immersed in some way with the work of Kant. As we will see in Part 2, this engagement with Kantian thought obtained enormous significance in both his lectures and his published work. It is important to keep in mind that his interest in Kant was to a large extent provoked by his resistance to Neo-Kantianism. As we saw in the preceding chapters, the site of contestation with the Neo-Kantians lay in their specification of the locus of truth in the proposition. The context for this location was that of an epistemological interpretation of Kant's *Critique of Pure Reason* which began and remained in the stems of sense and concept. As we have already ascertained, Heidegger is seeking to articulate a phenomenological reading of the philosophy of Kant that would be prerequisite to any consideration of epistemological questions. In addition to the texts which we will consider in Part 2, which concern themselves with the phenomenological question of the 'rooting' of understanding and sensibility in the transcendental imagination (original temporality), Heidegger also explicitly engaged in an analysis of Kant's notion or interpretation of being. He makes two significant references to 'Kant's thesis about being' in texts which span both his early and later writings, indicating that his fascination with Kant survived the so-called 'turn' (**Kehre**) in his thought. On the one hand, there is his 1962 essay of the same name, in which there is one of the few references Heidegger ever makes towards Marx. An earlier reference to 'Kant's Thesis About Being' occurs in *Basic Problems*, Chapter One, 'Kant's Thesis: Being Is Not a Real Predicate', and, it is explicitly connected with our discussions of original temporality as the source of a finite thinking of being-in-the-world. In his discussion, Heidegger considers two formulations of the thesis, negative and positive: 1) 'being is not a real predicate,' and 2) 'being is position or perception.' He does not equate these versions, as if they were two sides of the same coin. Indeed, Heidegger assents to the 'negative', but not to the 'positive' thesis, which he deems superficial, misleading and linked to Kant's suppression of the question of a common root in transcendental imagination, or, of an original temporality of existence. After an exploration of the two versions of the thesis, I will turn to the closing pages of the *Basic Problems* in 'The Kantian interpretation of being and the problematic of Temporality [Temporalitat]' in which Heidegger engages in a critical confrontation with the positive version of the Kantian thesis and proposes the necessity of a phenomenological self-interpretation and expression of Dasein from within the horizons of his own Temporal problematic. With his indication of the meaning of being as Praesens (and not the position of consciousness), we will explore the possibility of a mode of self-expression for Dasein which transcends the regime of predication, whether real, ideal or logical.

The negative version of the thesis: Being is not a real predicate

The negative version of the thesis, that being is not a real predicate, is appropriated by Heidegger as intimating, in an indirect way, one's existence, to the 'matters themselves', outside of a 'truth regime' of logic, 'ethics' and traditional ontology. Through an explication of the meaning of the 'negative thesis', we will be brought face to face with the problematic of the origination of 'basic concepts', or, of an indigenous conceptuality of existence. This is, among other things, the question of a 'justification' of Kant's derivation of his pure concepts from the 'logical function of judgements' and his positing of a *mathesis universalis* as the absolute context of Reality. This question is moreover another restatement of Heidegger's task of uncovering an 'ontological difference'.

Heidegger contends that the meaning of the 'negative thesis' rests upon the distinction in meaning between 'real' (**Realität**) and 'actual' (**Vorhandensein**, **Wirklich**). Real, for Kant, does not mean the 'actual', but is the 'what-content' of a concept of a possible object. For example, 'one hundred thalers', whether real or actual, has the same what-content, or, essential determination. Or, in a more fundamental case, existence cannot be a real predicate of 'God', a denial which is the basis for Kant's refutation of the ontological proof for the existence (extantness, actuality) of God. In this way, reality can make no claim with respect to actuality, or, to existence, in Heidegger's sense. Reality is distinct from actuality, as the 'what' is distinct from the 'how'. Moreover, each of these aspects is distinct from that which is most clearly expressed as this *existence* of the be-ing of oneself. In other words, actuality and existence fall outside of the 'relation' of predicative synthesis, of essential predication. Neither existence, nor actuality is a real predicate. Each adds nothing to the concept of a real, that is, 'possible object'. On the contrary, existence does not predicate or determine a 'thing', but, to stay for the moment only with the 'negative thesis', refers not to Reality, but 'beyond' the real, to a 'negation' of Reality.

It could be argued that with his negative thesis, Kant sets forth *by implication* a question of being, beyond the regime real predication. Yet, since, for Heidegger, reality and actuality are ontologically the same, an explicit questioning after being, which is guided by the notion of the ontological difference, is nowhere to be found in Kant. Heidegger admits however that 'seeds' of this difference may be found in Kant, but that these seeds remain dormant due to a failure to explore the question of 'basic concepts'. In this light, Kant's failure to uncover a difference betwixt reality, actuality *and* existence as the worldy being of the self, as Care, renders his work susceptible to the repetition of the violence of rationalism (as we can see in the example of most of the Marburg Neo-Kantians). Reality, for Kant, is determined by real predication, and is expressed as judgement. Insofar as reality is the totality of the real, the meaning of being in the context of the real is being present at hand, or theoretical being. With the Logical Positivists, this sense of concept (real) is fused with the empirical object (actual) as the criterion of application for the construction of knowledge. It will be Heidegger's task to show why these meanings do not disclose an appropriate sense of being for a phenomenology of lived, temporal existence.

The positive version of the thesis: Being is position/perception

Heidegger lays out the distinction between reality, actuality *and* existence, through an analysis of two senses of the word 'is'. On the one hand, we have the 'is' of the copula, of predicative synthesis, of judgement. The copula is a linking concept, an assertion of relation: something is posited relative to something else. Actuality, on the other hand, and, this is Kant's sense of the positive thesis, is instead a being-posited 'in and for itself', free of relation.[175] Actuality is further specified as 'absolute position', distinct from the 'real', which as possibility, is 'mere' position. An *actual* being is 'added' synthetically to its real concept in an existential synthesis, in which there is posited a relation between a thing and a concept. The predicative synthesis 'posits' real characteristics of a thing and its internal relations. This relation of 'reality' to an actual thing, as a synthesis of assertion, is termed, by Heidegger, 'full reality'. From the perspective of Kant's criticism of a mere play amid concepts, to posit the actuality of a thing, we must exceed the concept, go beyond its limits. Beyond the 'what' of the *res*, there is the 'more' of the positedness of the thing itself in the 'how' of its givenness. As Heidegger states, for Kant, this is not a predication, but an experiential concept, it is a 'synthesis of experience', a statement of actuality. Again, actuality is not a determination of the 'real' in terms of a 'what'-content, but of the 'how', as a 'way of being' of the thing in its 'actual' givenness.

Heidegger denies that there is a simple translation between the negative and the positive versions of the thesis. 'Positing' refers to the actual as given for 'my ego-state' as a relation of the thing and my thought of the thing, to the real, or, as the relation of the existent to the cognitive faculty. In this way, actuality, as a modality (Postulates of Empirical Thought), does not augment the concept but refers to empirical apprehension, to 'perception', sensibility. Indeed, he insists 'actuality' as absolute position is, for Kant, perception. Yet, immediately he indicates a catastrophic problem: if being is perception, then being is something 'subjective' that occurs in or as perception. Heidegger, in an apparent reference to Leibniz, exposes a monstrosity of a 'window with a perception attached . . .'[176] This 'monstrous site',[177] as we have seen, came to the foreground as the 'mythology of consciousness'. Heidegger, attempting, he alleges, to interpret Kant in a favourable light, indicates the monstrosity as a 'subject' bringing itself to a thing, perceivingly, in a 'relation that is aware of the thing'.[178] Yet, Heidegger immediately points out that the inadequacy of even this favourable interpretation in that perception is merely a being among beings (and one implicated under the discipline of reason). In other words, existence, be-ing, for Kant, is merely a being among beings. In and of itself, such a slippage back into the leading strings of theoretical and practical *reasons*, would be grounds for a dismissal of this thesis. Yet, Heidegger, as he will do with Kant's derivation of concepts in *Phenomenological Interpretation*, will seek to disclose the 'unsaid' or 'un-packed' in 'Kant', to 'understand' the thesis of Kant by immersing himself in it, seeking to follow its internal directionality for the sake of an interpretation that gives Kant the benefit of the doubt.

In *section b* of the chapter on Kant's thesis in *Basic Problems*, Heidegger breaks down his 'phenomenology of perception' into the elements of perceiving, perceived and being-perceived, or perceivedness, having already seen this latter in his *History*. As he is concerned with an expression which is 'outside' of real predication and statements of actuality, he passes over the perceiving and the perceived as candidates for the meaning of the thesis, as these are ontic aspects which are, for him, rooted in the basic orientation of being-in-the-world. Heidegger already fathoms, as the occasion for questioning has already been provoked by a uncanny disclosure of 'no-thing', that there is something more fundamental than a empiricist 'phenomenology' of the ordinary world, and of perception as a being 'in' the 'world'. Indeed, it would seem that the perceiving and the perceived are once again severed 'stems' disseminated from a basic existential rooting in the perceivability of phenomena. With this insight, Heidegger turns his attention to being-perceived, or perceivedness, as a clue to uncovering the positive sense of the 'thesis'.

Heidegger's hermeneutic of the tripartite perceived, perceiving and being-perceived serves to provisionally indicate *perceivedness* as an *a priori* 'ground' for finite thinking. The ground is a *being*-perceived, discoveredness (**Befindlichkeit**), disclosed as a temporal disposition amid being-in-the-world. Yet, even perceivedness remains dependent upon the given-ness of existence, upon the prerequisite *es gibt* of being-in-the-world. Indeed, the positive thesis, being is position/perception cannot of itself stand as an honest explication of this meaning of existence, 'Perception' is ambiguous; it begs for a rule, and thus, phenomenology becomes organized and disciplined by 'logic', by 'consciousness'. In this light, Kant's positive thesis becomes questionable since *being is not position*, or perception, but is disclosed through these modes of access (not to mention the fact that this tripartite remains subject to the positing act of a transcendental apperception). However, this does not necessitate that we simply 'throw the baby out with the bathwater'. Amid the play of *destruktion*, Kant's thesis points towards the deeper tapestry of intentionality, a *topos*, playspace of existence. Intentionality, existence, is disclosed amid the 'unity' of ecstatic temporality, and is not '. . . something objective nor something subjective in the traditional sense'.[179] Since a finite transcending has a directional sense, there must be an antecedent understanding-of-being in the moment of encounter in a temporal event.

The Kantian interpretation of Being and the temporal problematic

In the closing pages of *Basic Problems*, Heidegger critically engages Kant's positive thesis within the horizons of his own temporal problematic. He states that Kant's 'thesis about being' cannot, as provisionally suggested in the *History*, mean that being is *perceivedness*, in that such a meaning or sense presupposes an antecedent understanding, an illumination of be-ing. Amidst the transcending of finite being-in-the-world, there is a 'self-directing toward', which occurs as the 'clearing' of ecstatic-horizonal temporality, amid the 'unity of the phenomenon'. Existence, amidst its own original self-projection, lets beings be encountered in the opening which it has itself allowed to be. In its openness, Dasein receives the given, but in a way that is deeply 'spontaneous', instantaneous, suspended upon this temporal 'there' (**Da**). From out of these ecstasies, Heidegger concludes, erupts the directed transcendence which makes perception possible. An understanding-of-being which founds this 'directional sense' is made possible by means of a horizonal schema which orients the sense of an intentionality of perception. In this way, the horizon of *praesens* makes possible a perceiving of entities 'in the present'. Heidegger's 'un-said' version of the Kantian positive thesis projects an understanding of existence with respect to the horizonal schema of *praesens*. In this way, Heidegger comments, Kant's theoretical determination of the idea of being as position is 'grounded' originally upon the 'Greek' *hypokeimenon* (υποκειμενον) that has 'the character of ousia'.[180]

To further clarify the 'positive thesis' from the perspective of his own interpretation of Kant, Heidegger states that the thesis 'being is not a real predicate' must mean, for Kant, that being is a logical predicate. On his behalf, Heidegger refers to a posthumous manuscript of Kant:

> Accordingly, all concepts are predicates; however, they signify either things or their position: the former is a real predicate, the latter merely a logical predicate.[181]

In this light, concepts of position are 'identified' as logical predicates, interpreted in a temporal sense as a withdrawing of the being of the entity in the enpresenting of the entity. Heidegger states:

> In Temporal language, this means that a being can no doubt be found as extant in an enpresenting, but this enpresenting itself does not let the being of the extant entity be encountered as such.[182]

In the enpresenting, being is not encountered amid beings, but is understood in this enpresenting, and only in such an enpresenting. Being 'lets an entity stand', and *it* is understood in this standing, it is the *lethic* condition for the thing that arrives in the opening, and it is remembered in this opening.

As with any understanding, being is understood in this context as a projection, in the presencing of the thing in its self-determination. An ecstasis of original temporality allows beings to be encountered against a projective background of its horizonal schema. The temporal interpretation of an understanding-of-being unearths the limitations of Kant's notion of actuality as a *logical* predicate as it exhibits the sense of being of the proposition, predication or *logos*. Heidegger sets out his own alternative sense of being as the horizon of prae-sens,[183] already understood in each projection amid a temporal, lived existence. In each of the projections, being as the horizon, removes itself, projected as the horizon, to clear a playspace for this arrival of the entities which we encounter in everyday being-in-the-world. In this way, being is the 'prius', and makes itself heard amid the event of original temporality. It is here that Heidegger seeks to explicitly thematize the question of an 'ontological difference'. He laments that being has always been mis-interpreted as a being. Thales had said all was water, Kant, that being was perception, position. Yet, even in a mis-understanding, being becomes a problem, a question. Heidegger states, 'Since it exists, the Dasein understands being and comports itself toward beings.'[184] The distinction is latent in existence, 'it belongs to existence'. To exist is to be in an existence of ec-stasy, 'temporalized in the temporalizing of temporality'.[185] Our access to this distinction is 'founded' amid existence rooted in ecstatic temporality, *always on the outside of the cave*, in that the 'ontological difference' plays amidst this temporality, and thus, is uncovered amid our understanding-of-being of finite existence, as we exist 'in the truth'.

Towards an indication and expression of existence

Heidegger's *destruktion* of phenomenology in his *History of the Concept of Time* rejects the quick answer of 'consciousness' as the meaning of intentionality, setting forth in its stead an indication of the be-ing of the one who questions, of the questioning self, as Dasein, as existence. That which Heidegger retrieves in this *destruktion* is insight into the historicity, or the temporal singularity, of the meaning of one's being, this makeshift of human Dasein emerging from out of the opaque indifference of the Anyone and/or any-thing. Indeed, the problem of the being of the extant, and of a neglect of the question of being, is one and the same question of a difference of human Dasein from other things with respect to their respective beings. Heidegger assents to Kant's negative thesis, but contends that the 'outside' permitted by Kant, as a position or perception, does not truly transcend the being of the extant, does not go 'beyond' that 'reality', which is the domain of an analytical and synthetic judgement, and hence, of the *episteme* of logic. There must be an expression that indicates and shelters the phenomenon of lived existence, which means, as this is the temporal 'ground' of statement, a phenomenon which is pointed out and that in the face of which one must remain silent.

Bataille writes, in his much neglected work, *Theory of Religion*, 'Intimacy cannot be expressed discursively.'[186] He traces out a genealogy in which the archaic religions of the gift, of the 'sacred breach', become 'impossible' in the context of religions within 'the limits of reason alone'. Often in the works of Bataille, as with *Inner Experience* and *On Nietzsche*, for the sake of an expression of intimacy, he *shifts* into poetry. Such a strategem seeks to open up the field for indigenous expressions, to create a place for intimate expressions of the desired and of the impossible. Moreover, this expression raises the question of the possibility of 'originality' in philosophy.[187] Heidegger, who is rumoured to have read Bataille, and called him one of the most astute writers of France, invokes very similar limits to discursive expression in the form of real and actual predication. Unlike Bataille (who was not an academic), however, he did not enact explicit 'poetic expression' in the 1920s, restricting himself to *poiesis* of existential indication and to allusions and references to 'primitive self-interpretations of dasein' and other images, or metaphors, such as the 'common root'[188] and the 'Fable of Cura'. Yet, his lectures and writings themselves could, after all, be interpreted as expressions of his own indigenous *poiesis* (*creative* phenomenology) of lived existence. Existence is not a 'real predicate', 'it' exceeds 'laws of thought', of identity, contradiction, excluded middle, and, as with Schopenhauer's Will – the umbrella of sufficient reason. Existence violates, exceeds each 'law': it is 'strange', as an original temporality, history, it expresses the 'impossible'. He does not wish, however, to simply leave this thought off in some ambiguity, but seeks to clarify, 'beyond' the real predication of 'universal' and 'particular', this indication and expression of one's own existence. There are marked similarities between his own train of thought and that of Bataille. Heidegger, picking up the threads of Nietzsche's *Gay Science* in the opening lines of his *History*, indicates our predicament in the wake of the death of god, 'where' we are thrown to our abandoned singularity. Amid this 'absence of myth', this madman seeks to express his own 'being in the truth', as the ultimate myth. In an implosion of the severed 'stems' of logical predication, the madman finds his voice.

Dismantling the hegemony of logical expression

As we will explore in more detail in Chapters 10 and 11, Heidegger states that Leibniz is the founder of the principle of reason, who shifts the question of the 'essence of truth', in the last analysis, to a 'theory of judgment'. The locus of truth is the proposition which states 'something about something'. It can either be true or false. On the other hand, Heidegger alludes, a simple direct intuition, in this logical epistemology, is neither true nor false. For example, the proposition, 'The board is black' is a judgement of something about something. Yet, if Heidegger says or writes, board – black, simply pointing to this and that, without the copula of judgement, of a proposition, without grammar, mere juxtaposition, parataxis, then his logos is neither true nor false in Leibniz's logical sense. Such an expression transcends the state of affairs of logic and seeks to show through indication. We can detect again the latent presence of what Schalow named a 'rift between judgement and truth', a dissonance, which Schürmann indicates, is at the heart of truth as a-lethea. Dissonance points to not only the 'bifurcated intentionality', exhibited in the statement itself, as synthesis and diairesis (διαιρεσις), but also to one's own expression of being in the world in a non-syntactical pointing out, as in-dication. As we have seen, and in light of 'Kant's Thesis about Being,' there opens up a place for expression, transcending real and logical predication. Such an expression need not be propositional, it may be *poiesis* (or as grammatical in the sense of meaning as use in the 'language games' of the middle to late Wittgenstein).[189] We see such a concern for poetry in *Being and Time*, where Heidegger invokes the Myth of Cura, and in his later writings, where he explores the poetic thought of Sophocles, Holderlin, Trakl and Rilke. Yet, regardless of questions of expressive type, his overriding concern in the 1920s was the complex indication of a philosophical *logos* attuned with an expression of temporal existence. An analytic of dasein, a 'metaphysics of metaphysics', is an expression in this sense in that its underlying character is an indicative nexus of names that point out lived characters of temporality. Any philosophical expression could be interpreted as a formal indication of the 'matters themselves' if it read 'in the right way'. In this way, statement (*logos*) would remain tied to the matters and would not become free-floating logical nomenclatures, statements, judgements, propositions.

It is in the 'detours' of discourse, of transferable words, of 'interchangeable parts', that one's own original dis-closures of 'truth' are covered over, falsified and homogenized. Heidegger does not wish to nullify communication, but to contend that a real, logical and mathematical predication, in that each 'ensconce'[190] themselves away from, and over, existence, make an indigenous expression of finite existence impossible. The pathos of distance required by analytic forms of objectivity are incompatible with an intimate self-expression. They are, as Wittgenstein diagnoses, captivated by a particular grammar or more likely a form of expression, to one 'language game', when, in truth, innovation and difference is always possible if we seek to make ourselves understood to others. Heidegger insists in all his works of this period that 'theory' is merely one comportment, one way of expression, amid other expressions. The grounds of logic are not logical. That which Heidegger would call finite self-knowing is that which is open for its own possibility and expresses itself with these others in the openness of a temporal discourse, of our own fragility of meaning. It is here-there, where Heidegger's anti-Platonic currents come to the fore, in which he wishes to destroy a logic which, in a similar way to the pantheistic reason of 'Spinozism', seeks to create a monolithic world 'in its own image'. Heidegger attacks Kant in that he did not go far enough in his censure of pure reason, that his 'criticisms' were only a re-branding of reason, not its *destruktion*. An indigenous expression, on the contrary, would indicate that which is 'there', it allows a temporal disclosure, expression of 'matters themselves'.

In many ways, Heidegger's project is a radical fulfilment of Kant's criticism of the play amid mere concepts in the dogmas of determinism and 'rational' freedom. However, since Kant holds that truth is expressed, contained in the judgement, since each real and ideal predication is derived from the logical functions of judgement, we remain within a framework of pure reason, a disguised 'play among concepts'. Not bowing to the authority of reason, Heidegger insists that philosophy must transform itself, yet, 'transformation is only possible in seizing and maintaining what is essential'.[191] The 'sign' tells us to go to 'matters themselves'. The preparation of this ground and its cultivation is temporality articulating itself as the characters of existence. Real predication, on the contrary, is 'blind' to existence – it can only have scorn for this sketch of a *topos*, this 'there' of 'first philosophy'.

Part 2

The *Destruktion* of Ecstatic Temporality

*Agreement prevails only about pseudo-philosophy, but it is the agreement of the mob. Thus it would be a totally misguided conception of the essence of philosophy were one to believe he could distill **the** Kant and **the** Plato by cleverly calculating and balancing off all Kant interpretations. What would result would be something dead.*[192]

To a significant extent, as we have initially gathered, Kant has more relevance than Husserl for Heidegger's 'Sein und Zeit' project. It is Kant, who, however against his own philosophical proclivity, expresses the question of a tacit relation of understanding and temporality. Heidegger's engagement with Kant will be *destruktive* to the extent that he seeks to wrest free indications of original temporality from the Kantian architectonic. Yet, with Nietzsche, as every destruction is also a creation, I will examine parallels between the horizonal schema of the self-projection of original, ecstatic temporality in *Basic Problems* and the operation of schematism in the first *Critique* so as to cast some possible light upon the projected section on Kant in the 'Sein und Zeit' project. Part 2 will thus be concerned with *Kant and the Problem of Metaphysics* (1929), supplemented by specific references to Heidegger's lecture course, *Phenomenological Interpretation of Kant's Critique of Pure Reason* (1928). I will not only concern myself with a hermeneutic assessment of Heidegger's interpretation, but will also explicitly situate this interpretation within the phenomenological parameters of the 'Sein und Zeit' project. I will argue that the significance of Kant for Heidegger lies in his having set forth a rough sketch for original temporality in the transcendental power of imagination, the *a priori* root of the soul. Bringing Kant 'back into play', 're-worlding' him, Heidegger opens up a *playspace* for the 'translatability' betwixt Kantian thought and his own, which, it should be remembered, occurs as a 'recollection' from historicity.

Chapter 6

The Retrieval of Ecstatic Temporality

A normal response is to shrink back from the apparently absurd. Yet, if Kant's opening thesis demanding the unity of sensibility and concepts is warranted; if cognitional content can only come from sensibility; if the two 'stems' of cognition are operative only as one unified structure; if transcendental imagination functions only as it unites sense and thought; if transcendental imagination is the source of the schemata; and if these in turn define the limits of possible experience – if the conclusions elicited from the five stage inquiry are accepted – then Heidegger would appear to be merely pressing on to a necessary consequence.[193]

Heidegger's engagement with Kantian thought concerns the question of a 'relation' between thought and temporality. Kant is credited with being the first philosopher, however inadvertently, to have solicited the question. Heidegger writes in 'The Task of Destroying the History of Ontology':

> The first and only person who has gone any stretch of the way towards investigating the dimension of Temporality or has even let himself be drawn hither by the coercion of the phenomena themselves is Kant.[194]

At the climax of his 1927 lecture course, *Phenomenological Interpretation of Kant's Critique of Pure Reason*, Heidegger 'identifies' temporality and imagination with apperception. He rejects Kant's 'point of departure' in the severance of understanding and sensibility, one which restricts the meaning of 'time' to 'mere' inner sense, a linear, generic time. For Heidegger, Kant's conception of time fails to disclose the intrinsic relation between temporality and understanding. His very point of departure disallows, from the outset, any development of an indigenous conceptual expression of the phenomenon. In this light, Kant has constructed a boundary which forbids a *direct* relationship between apperception and temporality. Apperception as spontaneous 'act' *is* 'out of time'. Generic, 'linear' time, although it affects the concept of an object, and projects one of the conditions of possibility for experience, cannot 'touch' the timeless *a priori* of the citadel of spontaneity, asserted to be an empty, formal self-relation prior to 'experience'. For Kant, especially after the revisions of the *Critique of Pure Reason*, imagination becomes increasingly regarded as a mere 'middle', capable only of *figurative* synthesis. The question not only of the origin of concepts of the understanding, but also that of the source for the principles of reason, remains *unsaid*.

Heidegger, departing from Kant (and Aristotle) gives priority to pure, productive imagination as the formative centre of expressivity for an indigenous conceptuality. In *Being and Time*, imagination, as it was for Hamann, becomes the *topos* of language amidst the site of thrown projection, of the 'there' of mood and understanding which articulates itself as discourse (**Rede**). Kant refuses to admit a 'common root' in the transcendental imagination, instead interpreting schematism as a mere means of application, of *subsumption*, of 'unworlded' concepts onto manifolds of intuition. The pure interest which sustains this derivation of Kant's conceptuality from logical functions of judgement, and not from schematism, is his desire to maintain the autonomy and authority of 'reason' in its 'traditional' interpretation. Autonomy as distance, severance, is prefigured by a point of departure in the 'stems', these 'ascending gestures'[195] of sensibility and understanding. Heidegger attempts to destroy, dismantle, that which in Kant was 'kept on' merely to follow a 'party' or tradition, that which does not allow an original phenomenon to express itself. In this light, it could be argued that Heidegger radicalizes Kant's criticism of a play among 'mere concepts', charging him with a lack of sufficient radicality.

With the displacement of transcendental imagination, it was impossible for Kant to witness the intimacy betwixt temporality and apperception, and to fathom original temporality as a 'call'[196] for self -articulation, for an expressive discourse of existence. The emphasis upon the 'obscurity' of the Kantian 'doctrine of the schematism' brings the transcendental power of imagination into the heart and soul of *Being and Time*, an entry which can shed light upon the temporal character of resoluteness.[197] Heidegger charges that Kant shrinks back from the chance for an encounter with the obscurity of imagination, an 'art hidden in the depths of the human soul'. In his evasion, Kant suppresses his own temporal predicament with 'covert judgments' of the 'common reason'.[198] Heidegger contends: '. . . the decisive *connection* between *time* and the *"I think"* was shrouded in utter darkness; it did not even become a problem.'[199]

Heidegger's detour through Kant

In the following, we will travel through Heidegger's footsteps in his phenomenological reading of Kant. This reading is expressed within the parameters of Heidegger's 'idea' of fundamental ontology. This 'ontology' does not seek to re-inscribe the Kantian architectonic back into the closed circle of the play amid mere concepts, the Platonic flight of the dove, but to take him at his word, *to take him to task* with respect to his criticism of *pure* reason. Moreover, in that Heidegger himself admits that his reading of Kant exceeds the latter's own self-understanding of his work, this reading takes him beyond his word, in the destructive exposure of 'pure reason' as a pure, *sensible* (temporal) reason. Despite this admission, he would be the last to suggest that his work is thus irrelevant to an interpretation of 'Kant'. Moreover, as Sherover and others have argued, it would be a distortion of the facticity of the Kantian text to merely regard it as a fixed and known quantity, as if it had fallen out of the sky as a final revelation. In this light, the Kantian text remains a place of contestation; its 'meaning' is not fixed.

At the same time, there persist obvious differences between Kant and Heidegger, not only in that Heidegger's interpretation of Kant is motivated via his own project of radical phenomenology, but also due to the limitations inherent in the structure of Critical philosophy. Henrich himself admits that for Kant any question of the 'Being of the self' must remain mute.[200] There are questions that simply cannot be dealt with by Kant insofar as the intention and articulation of his work privileges, as I have suggested, a derivation of the pure concepts from the 'logical functions of judgment', and thus, a *subsumptive*, and not an *expressive* understanding of knowing – through which intuition, sense, imagination are contained, put to work 'within the limits of reason alone'.

In his phenomenological reading, Heidegger points to the places of ambiguity and tension within Kant's first *Critique*, and to those places where what is said seems to suggest confusion, a 'disquieting unsaid', where lowly origins begin to show through the cracks, where the hegemony of conceptual logic begins to break. Heidegger points to these places where Kant is susceptible to a re-interpretation with respect to the possibility of the pre-theoretical and pre-practical expressivity of existence. Heidegger, as he admits throughout his works, is not serious about being a 'Kant scholar'. Neither content to be merely a critic nor to remain within the language of the *Critique*, he must wander to a significant extent on the 'outside'. At the same time, no one can deny to Heidegger a vast knowledge of Kant, and a keen interest in 'getting it right', or, his statement that an 'external' criticism would always be unjust. Indeed, from the 'beginning', and especially in the German Idealist movement, there have been many who have sought to out-Kant Kant. Heidegger contends however that one must transcend the Kantian framework if there is to be any clarification of the relation between thought and original temporality. In his re-writing of Transcendental Philosophy, Heidegger will push Kant beyond the limits of *his own* intelligibility.

The primary direction of *Kant and the Problem of Metaphysics*, Heidegger's first published work after *Being and Time*, is determined by a formal indication of a rooting of the 'stems' of sensibility and understanding in the transcendental power of imagination, or, as he admits, even in this work, 'original time'[201] (**ursprüngliche Zeit**). Heidegger traces finite thought, *philosophical conceptuality*, to Dasein, the existent self in the temporalizing of its own ecstatic temporality. Suspending the question of Kant's *Deduction*, of the possibility of relation of the two stems, Heidegger points to an intimate self-openness amidst an 'event' of disclosure, self-interpretation, and expression. The 'stems' are merely, as Sherover remarks in his work *Heidegger, Kant, and Time*, aspects, abstracted from the pre-experiential awareness of the self. Heidegger questions Kant as to the possibility of an indigenous site of a finite knowing amid the temporal horizon for a disclosure of *being-there* existence. The impossibility of this question for Kant is reflected in Cassirer's critical review of *Kant and the Problem of Metaphysics*,[202] in which he warns that Heidegger's radical temporalization of thought would deny to Kant the object of his desire, which is eternity. Kant was merely a 'critical theologian', as Nietzsche would have it.

Ecstatic temporality and the problem of metaphysics

The problem of metaphysics, as laid out by Heidegger in *Kant and the Problem of Metaphysics*, is, for Kant, the question of the possibility of *a priori* synthetic judgements. What Kant is seeking, in his paradoxical formulation, is a type of knowledge which exists originally, prior to empirical experience, as the synthetic unification of concept and pure intuition. We begin with experience and seek its conditions of possibility. Yet, since synthesis is the '. . . mere result of the blind power of imagination', and since this is to be an *a priori* synthesis which is necessary for there to be knowledge, there must be, as Heidegger contends, with support from Kant's text, a pure, productive or transcendental, *a priori* imagination, or, as we have seen, a transcendental power of imagination (**Einbildungskraft**). It is such a solution to the problem of metaphysics that is most resisted by critical interpreters of Heidegger, such as Cassirer[203] and Henrich.[204] Each denies that Kant's question, 'How are synthetic, *a priori* judgments possible?' finds an exclusive answer in transcendental imagination, as the 'third basic faculty of the soul'. Henrich alludes to Cassirer's assent to a rootedness of thought in imagination,[205] yet, both stood shoulder to shoulder in their opposition to the radical implications of such a rootedness as it comes to light in Heidegger's hands. Henrich and Cassirer are careful to point out the intractable difficulties that such an interpretation poses for the integrity of the 'Kantian project'. The primary issue at the end of the day is the 'authority of reason', of its logical detachment (the sense of its interpretation of the *a priori*) unto itself, secure from 'time' detachable ultimately and necessarily from intuition and all acts of understanding, and all objects of synthetic conceptual-intuitive operations. It is the 'logical' (as with the neo-Kantians) interpretation of Kant's *thesis about being* which lays the ground for the *pathos* of reason. If 'reason', understanding or pure concepts could be traced to a transcendental power of imagination, such autonomy, distance, 'unliving', as expressed in one of Heidegger's early denunciations, will no longer be able to stand. In other words, thought would be implicated amid the horizon of original temporality. The implications of such a reversal, or 'turn of events' in the historicity of thought, would be that be-ing can express itself intimately amidst existence, from out of this *topos* of an indigenous, lived historicality.

Such gestures of intimacy, entailed by Heidegger's radical phenomenology, would seem, for Kant, to be, ironically, infringements upon the autonomy of the person and the authority and dignity of 'reason'. Such intimacy would annihilate the very possibility of a *strict, universal, necessary* knowledge as it would subvert the rationality of his differentiation between *a posteriori* and *a priori* knowledge. The peculiar significance of what he phrased as the 'Copernican Revolution' is in fact at stake: the possibility of an *a priori* synthetic knowledge which relies upon the 'turn to the subject'. Knowledge will no longer follow the leading strings of nature, keeping us in a state of blind groping, darkness and ignorance. No longer will the criterion of *a priori* knowledge be defined as the adequation of thought with an object, but in reverse, of an object with thought.

Heidegger's suggestion of a pure, *sensible* reason, seen against this background, would not be merely a contradiction in terms for the Kantian, but a counter-revolutionary attack against the triumphant beast of *Enlightened Reason*. Heidegger is indeed, as we have seen in his description of phenomenology, advocating a return to the *object*. Yet, as he has suspended the representational apparatus, his 'Greek' sense of the object would not be that of either the theoretical 'concept of an object' or the empirical object in the first *Critique*. He invokes a pre-theoretical dimension, which is pre-objective, and thus, in Kantian terms, would be more properly a realm of no-thing, intuition, imaginary being (**ens imaginarium**), that which for Kant can never give rise to universal and necessary judgements, propositions, knowledge *in the strict sense*. For Kant, the question of such a 'unity', of such a *being*, must remain mute.

Laying the ground for the problem of metaphysics

What is most striking about the Kantbook, in its posturing as **the** interpretation of Kant, is its radical novelty of expression. Immediately, we gather that this is not Kantian 'scholarship' as usual. Except for the occasional quotation, the primary vestiges of Kantianism have been entirely jettisoned. In other words, *it does not sound like Kant*, although Heidegger, not conceding rigour, is at pains to deploy Kant as the *logos* of his own phenomenological indication. We have the sense, from the outset that there is 'something strange', peculiar about his analysis of Kant. One senses that there is occurring a full-scale procedure of *translation* – a lexical shift is being made from the Kantian problematic, that of the question of the 'possibility of a priori synthetic knowledge', towards a house of a differing language possessed by the question of an 'inner possibility' of ontological knowledge'. This shift intimates the transfiguration of the meaning of 'interpretation' in light of the temporal problematic.

The Kantbook does not promise to be an interpretation of Kant *in the usual sense*, but an attempt to illuminate via the 'idea' of fundamental ontology an interpretation of the first *Critique* as 'laying of the ground for metaphysics'. What is being tested or disclosed in this exercise, in Heidegger's active phrase, in this 'performance', is the formal indication of a pre-theoretical/practical existence. Amid this 'origin', which originally 'gives forth' conceptuality, there is the possibility of a different expression. Heidegger's deconstructive method, declared in *Being and Time*, is so blatant that one wonders why his Kantbook has provoked such intense hostility from the 'guardians' of the 'real' Kant.

The problem of metaphysics is a question of the possibility of pure horizons of discernability, of finite knowing which in its being is a comporting 'relation' *a priori*. Such an intentional relationship opens up a playspace for the experiencibility of beings – or, as with the *Phenomenological Interpretation*, a pure turning-towards that which is given from itself, amidst a pure 'object relatedness'. With such a situation, finite knowing is disclosed as a *pre-aesthetic syndosis of imagination* which is the source for the *a priori* unification of the aesthetical elements of space and time. The formal indication of existence is one's anticipation of that which is sought, and not, in reference to Kant's self-description, of an 'uprooting' and 'dissection' of a field. At the same time, a point of departure in Kant is necessary since the possibility of self-interpretation '. . . is not yet and no longer in possession of the original directive force of the projecting,'[206] of a projection of an inner possibility of *a priori* self-knowing. There must be a historicity of everyday orientation, of being-with (**Mitsein**), a factical fundament. The projection of the indication of existence anticipates and guides as that which is forecast prior to the 'carrying out of the ground laying'.

Ground-laying, for Heidegger, intimates the 'essence of knowledge in general', a projection and retrieval that casts forth a horizon of possibilities for an excavation of intuition and thought to the transcendental power of imagination. The essence of 'knowledge in general' and the 'tactical limitedness of knowledge' are disclosed from out of the horizons of human finitude, intimating the primacy of intuition. Heidegger writes:

> In order to understand the *Critique of Pure Reason* this point must be hammered in, so to speak: knowing is primarily intuiting. From this it at once becomes clear that the new interpretation of knowledge as judging (thinking) violates the decisive sense of the Kantian problem.[207]

Heidegger, following Husserl, specifies finite knowledge as 'primarily an intuiting'. Finite Knowing is not a judging.[208] This contention incites the battle over an ambiguous text and Heidegger's claim to have disclosed the 'Kantian' problem. He describes this field of original thinking and intuiting as the 'sufficiency' of lived existence. He writes,

> All thinking is merely in the service of intuition. Thinking is not simply alongside intuition, 'also' at hand; but rather, according to its own inherent

structure, it serves that to which intuition is primarily and constantly directed.[209]

An 'inherent relationship' is possible since intuition and thought are each a constituent of knowing, itself a type of intuition, with each possessing capacities for receptivity and spontaneity. In this way, intuition, in the strict sense, has priority over conceptual, logical thought, in that sense has an immediate relation to the phenomena while thought has a mediate, or discursive, relation with the phenomena. He writes: 'The circuitousness [Umwegigkeit] which belongs to the essence of understanding is the sharpest index of its finitude.'[210] Both are intuition, yet, there is a difference, one that arises from the severance of thought and being, from the abstraction of the concept from existence, just as Nietzsche wrote of concepts in his *On Truth and Lying in the Extra-Moral Sense*, as 'worn-out metaphors'. The concept has, to use Sherover's term, been 'de-temporalized'.[211] It becomes 'free floating', in a derogatory sense, since loose concepts can be applied to non-indigenous, non-appropriate situations. There persists, nevertheless, for Heidegger, an 'inherent relationship' betwixt sense and concept, as each is a comporting, a possibility or 'way of being'. Without a reciprocal 'joining' of these elements, there would not be any knowledge. Yet, with phenomenology, intuition remains 'first' the 'essence of knowledge in general'; thought is a conditioned consequence of our own finitude, it acts in order to indicate phenomena. Thinking is a 'mark' of finitude; a divine knowing (*intuitius originarius*) would consist only in intuiting, it would bring the object of its intuiting into being through its own intuiting, and thus, '. . . it cannot require thinking.'[212] It is the finite creature who requires thinking, one that *should be* in service to intuition, as would occur with a pure, *sensible* reason, thought creatively receptive amidst finite transcendence.

The essence of the finitude of knowledge, as we have seen, is, for Heidegger, indicated through its dependence upon an 'object' being given. That which indicates the essence of the finitude of intuition is thus its receptivity. The being-there of knowledge is finite, it has been given over to a place of beings, a circumstance in which a finite knowing 'takes things in stride'.[213] Heidegger specifies this receptivity in his contention that Kant 'for the first time attains a concept of sensibility that is ontological rather than sensualistic'.[214] In other words, receptivity is not concerned with appearances, with sensation, but with the things themselves, with being. For Heidegger, Kant's *a priori* sensibility is to be taken as an indication of his own indigenous 'conceptuality', 'read off' or expressed with respect to this phenomenon which is 'there', prior to theoretical representation and questions of apodeictic certainty. 'Theory', for Heidegger, can only have an indicative significance. Expression is merely pointing to a 'thing': finite knowing 'takes a look' at 'the general' so as to 'direct itself' to that which is 'there'. A concept is a 'representation of a representation', and thus, is a predication, and temporally, a premonition, 'an assertion of something about something'. Judgement, in it relation to an object, is dependent upon intuition, united to it through the synthesis of imagination. The 'stems', each to its own, can not make a knowing, but only in *synthesis*, which is also the *modus operendi* of judgement, is this finite knowing fulfilled as the veriative synthesis of 'thinking intuition'. For Heidegger, this synthesis 'breaks open' an original dimension 'where' be-ing can be encountered and expressed.

An infinite 'knowing' creates its own object. A finite knowing '. . . alone is delivered over to the being which already is'.[215] Finite knowing moves towards these 'matters themselves', but, it 'conceals at the same time and it conceals in advance . . .'[216] It is thus forbidden any 'relation to an infinite'. But, the 'unity' of the stems, the elements, 'is no mere collision'. Yet, 'what united them, this "synthesis," must let the elements in their belonging together and their oneness spring forth.'[217] Synthesis is the ground of an 'original unity'. Kant does not pursue the question of the common root, and as a consequence, he fails to unfold a laying of the ground for metaphysics as an excavation of a 'first' (**zunachst**). Heidegger asks:

> How can a finite creature, which as such is delivered over to beings and is directed by the taking-in-stride of these same beings, know, i.e., intuit, prior to all [instances of] taking the being in stride, without being its 'creator'?[218]

The path to a 'first', which, in the context of the Kantian problematic, 'leads to the unknown',[219] is an exploration of concealed landscapes, caught in the web of representation. 'The uncovering', Heidegger describes, is an "unknotting" as a freeing which loosens the seeds [Keime] of ontology. It unveils those conditions from which an ontology as a whole is allowed to sprout'.[220] Heidegger has jettisoned 'Kant' for the 'inner movement of the Kantian groundlaying . . .'[221] His unknotting is a breaking, as a plant breaks out of its entombment in concrete, that announces pure imagination as the 'formative center' of finite self-knowing, in which a spontaneous receptivity allows the phenomenon to 'show itself from itself'.

Chapter 7

The Excavation of Ecstatic Temporality

Kant for the first time came upon this primordial productivity of the 'subject' in his doctrine of transcendental productive imagination. He did not succeed, of course, in evaluating this knowledge in its radical consequences, by which he would have had to, as it were, raze his own building, with the help of the new insight. On the contrary, this great intuition was, in principle, lost. Nevertheless, this first advance into the transcendental imagination, which was for Kant only obscurely connected with time, was the first moment in the history of philosophy in which metaphysics endeavored to liberate itself from logic, and from a logic which had not found and never did find its own essence in metaphysics, but remained a training grown superficial and formalistic. Perhaps the true happening in the history of philosophy is always but a temporalization [Zeitigung] of such 'moments' in distant intervals and strokes, moments which never become manifest as what they really are.[222]

It is not sufficient to simply unfold *pure* imagination as the root of the 'stems', or to contend that Kant was the 'first' and 'only' philosopher of the tradition to intimate the question of a specific interface between thought and temporality. On the contrary, this would be to stay within the Kantian framework, remaining with the stems and presupposing these. Instead, we must attempt to surpass this historical and artificial point of departure, and, through the razing of the Kantian 'building', grasp that the schematism of the imagination, as a making-manifest of concepts, itself 'projects' its own expression, a conceptuality of finite knowing, and in this way, is another indication for ecstatic, original temporality. It is interesting that Heidegger sees in Kant answers and intimations to the deeper question of the meaning of *Being*, and thus, is able not only to criticize him, but to surpass him as well. In this way, he is not seeking, as I have alluded, to be a 'good Kant scholar', but draws upon him and others for the sake of a retrieval of the question of temporality as the transcendental horizon of being. In the following pages, we will trace the 'descent' from the 'stems' to an 'existential rooting' of sense and concept in the schematism or the transcendental power of imagination. This translation of this latter power into original, ecstatic temporality is indicative of Heidegger's *destruktion* of the history of ontology, and his own attempt to excavate temporality as the transcendental horizon for the question of being.

The descent from the stems: Kant and the problem of metaphysics

In his re-articulation of transcendental philosophy in *Kant and the Problem of Metaphysics*, Heidegger enacts a deconstruction of the first half of the First *Critique*. He travels along with Kant on the pathway of the Transcendental Aesthetic and Analytic. But, when he reaches the Schematism, we are strangely turned around and are made to proceed backwards. It is in this way, however, that we are able to see the 'origin' of thought in the transcendental imagination. We will begin with the aesthetic.

With respect to finite knowing, and remembering the phenomenological character of Heidegger's readings of Kant, pure intuition, space and time, is 'first', an *a priori* condition of experience, or in Kant's language, an ideality of the 'mind'. Space and time are at once forms of intuition and formal intuitions as each can be itself intuited in pure intuition immediately and as a whole. In terms of the former, space and time are projected as the forms of intuition in the constitution of experience. In regard to the latter, the many spaces and times are limitations of a unique space and unique time. For Heidegger, bringing this consideration of pure intuition into connection with Kant's 'thesis about being', pure intuition intimates many intuitions without being a real predicate. It is not only a specific possibility of re-presentation for that which is represented, but also each of its 'parts' contain the 'whole' as a formal intuition. Heidegger writes, 'It precedes all the parts as the limitable, unified whole'.[223]

Time is a non-spatial re-presentation of 'inner sense' which, as receptive, is a 'succession of states of our mind, representations, drives, and moods'.[224] Time is experienced as 'pure succession', and it is, for Heidegger, an unobjective, un-thematic self-intuiting. Time for 'Kant' (and Heidegger in a differing way) is pre-eminent over space in that it is, as the most subjective 'form of intuition', the basic, '*a priori* condition of all appearances whatsoever'.[225] Since all representations are manifest within time, those of space, of outer sense must fall 'in' time, if only mediately to be 'taken up' into an 'inner sense'. Time 'dwells in the subject in a more original way than space'.[226] Time exhibits an intimacy with the self. At the same time, the self, as ecstatic, projects an existential spatiality as the *topos* of encounter. Time, for Heidegger, is a playspace of intuiting grounded in the 'belonging together' of finite existence, it is an intuiting which 'allows the encounter'.[227] Time is thus the basic horizon of the lived existence of the self, through which disclosures of beings are possible. Heidegger writes:

> Time reduced to the givens of inner sense . . . is at the same time only ontologically more universal if the subjectivity of the subject exists in the openness for the being. The more subjective time is, the more original and extensive is the expansiveness (Entschrankung) of the subject.[228]

If the subject does not exist in the 'openness for the being', such a time would thus not be 'ontologically' more universal. As I have indicated, common time, dominated by theoretical 'objectivity', covers over a 'more subjective time . . .' An 'objective' time reduces one's self to a point, to an Anyone (**Das Man**). Openness, as indicated by Heidegger, is, contrary to the reservations of a Husserl or Kant, an openness amid being-in-a-world, which is neither linear nor logical, but, is sur-linear and sur-logical, as an original temporality that expresses itself. Yet, as we have been thrown into the web of antithetical 'stems', such an indigenous self-expression, in terms of Kant's problematic in the *Critique*, is suppressed, covered up, in the uprooted conceptuality of pure thought.

Heidegger turns to the Transcendental Analytic in an attempt to excavate the *topos* of indigenous self-expression from beneath Kant's architectonic regime of subsumptive conceptuality. He begins with the claim that pure thinking, the other stem of knowledge, is, for 'Kant', a 'determinative representing'. It is directed towards intuition, but in service to the phenomenon alone. An object (**Objekt**), projected in a representation, is a particular and is determined as such, with respect to the 'universal representation', by which one concept 'applies' to many. General concepts, however, in their 'averageness', cannot 'hit upon its original essence', and thus, the situation is dominated by the intentions of an average expression, and of its subsumptive procedure. Instead of an apprehension of the singular, as a pointing to *this*, there is an agreement of the 'many' in the one (**das Eine**). It is here where that the question of expression finds itself at a crossroads. There is the danger that we will forget the character of conceptual representation as a determining act in the service of intuition and undertake a reversal in which the matters are subsumed by the concept in terms of its mere unworlded form. Heidegger, as we have seen, sets out an alternative interpretation of conceptual unity in a description of the features of an average knowing. These aspects will guide his own articulation of finite knowing of an *eigentlich* self. He writes: 'The oneness of this one must be anticipatively kept in view in conceptual representing . . .'[229] Such a projection of oneness is a 'preliminary representing of unity'[230] where a representation is formed into a 'reflected concept'. The representation is gathered into a 'unity' amid a precursory 'look' (as with εἰδος as the shape of a thing). This precursory look, or the projective indication of oneness, arises, Heidegger contends, only amid the playspace of reflection, and 'guides the unification'. Understanding is thus an 'original comprehension of grasping', and the represented unities 'are the content [**Inhalt**] of the pure concepts'.[231] Or, 'The pure understanding itself yields a manifold, the pure unities of possible unification.'[232] In that 'unity' itself is the content of the concept, Heidegger places the traditional conceptual form into question:

> The content [Wasgehalt] of these concepts is the unity which in each case makes a unification possible. The representing of these unities is in itself already conceptual *a priori* on the grounds of its specific content [Inhalts].

> The pure concept no longer need be endowed with a conceptual form; it is itself this form in an original sense.[233]

It is significant that Heidegger is offering a differing account of the *pure* concept as that which has 'unity' or 'unifying' as its specific content, in the sense of an 'act'. Such an act is an openness to the phenomenon, its precursory *look* of oneness allows for the self-showing of the phenomenon. Moreover, this *look* is of itself guided by the phenomenon and could be said to be the self-expression of the phenomenon in its unity. Yet, so as to understand the possibility of a *nonconceptual form*, we must be cautious of the distortions that arise amidst the artificiality of the stems, and which has necessitated that Heidegger give a differing account of the pure concept. We already know that a formal concept cannot account for this 'unity' of the phenomenon. We also know that a transcendental logic that is constrained by the laws of logic and the authority of reason will only be capable of producing a subsumptive relation between the particular and the universal (as one interpretation of the schematism). In this way, each of these logical possibilities nullifies the self-expression of the phenomenon in its singularity. Heidegger, in the current reading, traces the temporal unity of the phenomenon through a destructive engagement with the 'stems' guided by an anticipation of the 'emergence' of temporal concepts from the schematism. This implies that the sense of a 'concept', for Heidegger, refers to his own indication of an understanding-of-being, which, we could argue, has a much greater kinship to a pure, productive imagination than to Kant's mere 'faculty of rules'. Understanding, in Heidegger's sense, is 'there' amidst the dispositions of anxiety, boredom, in 'truth situations' that are themselves 'conceptual' in a phenomenological (indicative) sense. In other words, pure concepts are not species of 'reflection', but *a priori* acts of unifying, 'reflecting concepts' which are inherent in the structure of reflection. The original *topos* of concept formation, this understanding-of-being, pure imagination, is a making manifest of ways of being, looks, expressions of 'unity' (or, *a priori* horizons for world entry, as in *The Metaphysical Foundations of Logic*). It would lead to distortion, we must remember, if we were to consider the 'origin' of pure concepts only in the context of the severance of the stems. To remain merely an expositor, a critic of Kant, would be to miss the decisive problem, that of *original* temporality, to remain lost in a house of mirrors but without questioning and deconstructing this severance, and subsequent, entanglement of 'stems'.

The essential unity of pure knowledge

The initial violence/ruin of the theoretical severance of the stems, in its distortion of an original coherence or intimacy of intuition and expression amid lived existence, cannot be subdued through a mere mediation, as with Aristotle's 'supervenient' joining or linking in his *De Anima*. Such mediation would only deepen the distortion. On the contrary, as with Schelling, a reciprocal dependency of the stems of knowing indicates an original situation that not only gives rise to these stems, but also 'maintains them in their unity'.[234] According to Heidegger, such an original unity is 'manifest' in Kant in the question of the essential unity of pure finite knowledge in the 'key' section of the First *Critique*, 'On the Pure Concepts of the Understanding, or the Categories.' Each concept, despite its historically conditioned free-floating status, intimates the 'traces' of 'unity', and, each is involved in the synthesis of imagination operative in the constitution of knowledge. The original 'unity' of the stems, a 'unity' which each presupposes in its synthetic capacity, must have a 'superior' capacity of projecting syntheses, or 'forming' these syntheses amid their unifying. It is this question of an original source that discloses the trace of the 'unity' of root and stems in the pure imagination. Heidegger underlines his problem with Kant's place of departure in the First *Critique*:

> The question concerning the essential unity of pure intuition and pure thinking is a consequence of the previous isolation of these elements.[235]

The path back to the origin is an excavation, which undertakes a *destruktion* of the stems, a methodical dismantling, shaking up, a re-configuration of the meaning of the Kantian elements, for the sake of an uncovering of 'how each of these elements structurally supports the other'.[236] With transcendental logic, as distinct from general logic, a synthesis of representations is presented to the concept, which requires the other, aesthetic, as it would be empty without this *materia*. The necessity of synthesis exposes the receptivity of intuition, which must 'always affect the concept of the object'.[237] The spontaneity of our thought must wait for a synthesis to occur, even in its affectation and anticipation, it is dependent upon receptivity. Heidegger, underlining the finitude of our knowing, writes that 'our pure thinking always stands before the time which approaches it.'[238] Pure thinking and pure intuition celebrate, as Nietzsche writes in *Beyond Good and Evil*, their 'marriage of light and darkness'. Yet, Heidegger describes less of a marriage than a situation of temporary liaisons, as events of thought. He writes:

> Both pure elements come together from themselves from time to time; it joins together the seams allotted to each, and so it constitutes the essential unity of pure knowledge.[239]

Pure synthesis enacts these liaisons, in the first instance, as a 'mediating'[240] of the stems, which contain the 'basic character of the two elements'. Pure synthesis is, Heidegger traces, an act that occurs as *syndosis*, *synopsis* and *synthesis* amid the orchestration of finite knowing. Pure synthesis is thus the 'root' of unity for finite knowing, in that it projects a guiding thread for these traces of unity detected in the stems. He contends that the First *Critique*, as a 'laying of this ground for metaphysics', reveals pure synthesis as the origin of pure knowledge, even if in this provisional indication, the matters remains 'deeply veiled'. This 'unity', he writes, cannot be a creature of mediation, but is instead a rootedness: 'this center is a structural one,' 'joining into one' as an 'original, rich wholeness of one composed of many members'. Henrich has described this 'wholeness' as a 'unity' of 'equiprimordial moments', as a 'unifying and giving of unity.'[241] 'Unity' is a horizon of belonging together which incorporates, in a veiled way, the three syntheses. In this light, pure knowledge is a 'multiform action which remains obscure'. As a comportment of existence, it can only be understood amid its performance, event, '. . . traced out in its springing-forth'.[242]

As an alternative to the concept of the severed stems, Heidegger, as with his differing account of the pure concepot, sets out the *category* as distinct from, and displacing, both sense and concept. The category, in the sense of Heidegger's description, is not a schema of speech (σχημα τος λόγου), judgement, proposition, but a schema of being (σχημα τος εντος). 'Category' in this sense, is not an empty logical form isolated in itself, but includes in its own essence the original unity. In this way, an understanding-of-being 'lies in it directly' as the 'pure intuitability of the notions'.[243] The category is thus consistent with Heidegger's earlier description of the content (**Wasgehalt**) of the concept being unity itself, and concepts as unifying unities which had transcended the necessity of conceptual form. A category (**existentiale** in *Being and Time*), abides the seeds of this original unity and cannot be reduced to either 'stem' in isolation. It is neither of the 'Transcendental Aesthetic, nor the Transcendental Logic'.[244] Heidegger writes that the question of the method to ex-cavate this originary 'middle' remains 'foreign to Kant'. In his preference for logic over aesthetic, Kant is blind to 'matters themselves'. The logical and architectonic control of the 'complicated web of human knowledge'[245] prohibits, Heidegger declares, the question of the original unity of pure knowledge. He writes that a deconstruction of the Kantian architectonic seeks to disclose original 'unity' beneath these collateral 'ruins'.[246] An ex-cavation of original, ecstatic temporality must 'break through the architectonic of the extrinsic succession and pattern of problems'.[247]

The inner possibility of essential unity

If we listen to Heidegger's statements that the descent from the stems to the jointure of the trunk has been exposed as a mere accessory, then the epistemological question of the relationship of sense and concept is uncovered as a derivative 'problematic' that has somehow asserted itself as primary protocol.

If this is not the appropriate place to *begin*, then there must be some *elsewhere* where we can disclose and articulate the unity of existence *before* the stems. Heidegger, even as he wanders through his Kantian detour, has, with his *destrucktion*, cleared a *topos* for an intimate self-interpretation of existence, or, an 'original self-forming of an essential unity of ontological knowledge'.[248] This original self-forming is an 'illumination of the inner possibility of finite knowledge'.[249] Heidegger contends that it is the analytic of Dasein, as a fundamental ontology, which uniquely indicates the *topos* of lived existence. Finite knowing emerges amidst a pre-understanding, a matrix of comportments, orientation – as a being-by beings before and beyond the influence of abstractive and analytical strategies of interpretation.

Finite creatures exhibit a basic trait of 'turning-toward' amidst lived existence, of projecting the horizons for object-relatedness, for that which stands against (**Gegenstand**). In this turning-towards, the finite creature first allows a space for play (**Spielraum**) within which something can 'correspond' to it. To hold oneself in advance in such a playspace, to form it originally, is, for Heidegger, none other that the transcendence that marks all finite comportment to beings.[250] This *topos*, place of play itself is *where* existence knows. Finite knowing 'takes in stride', it is receptive, it cannot create a thing, it is not able, from out of itself, to 'place the being before itself'.[251] In this way, for Heidegger, finite knowing, as an original *a priori* comportment, is a turning-towards that which offers itself. Turning-towards exhibits the intentionality of 'ontological' projection, in that finite receptivity, with respect to the be-ing of the being, must have a 'faculty of letting-stand-against-of', one 'from out of ourselves' that opens up a playspace, in which we must hold ourselves in advance. This playspace is a 'no-thing' (**ein Nichts**), with respect to that which exists as a being. In this way, finite knowing projects its own horizons of existence, but at the same time, is suspended in the midst of phenomena which stand-against (**als gegenstehendes**), 'there' with their own self-expression, 'free' from our control. Play points to creativity – attempts to indicate, express 'it' – and thus, will be provisional, makeshift.

> Only if the letting-stand-against of . . . is a holding onself in the nothing can the representing allow a not-nothing [ein nich-Nichts], i.e., something like a being if such a thing shows itself empirically, to be encountered instead of and within the nothing.[252]

The nothing offers 'the possibility of a preliminary Being-oriented toward the object'. Or, as Heidegger punned in his Inaugural Address at Freiburg, 'the nothing nothings', or the 'naught naughts', amidst an 'ontological turning-toward', or, as expressed in the language of the *Phenomenological Interpretation*, as a projection of a horizon for the objectifying of 'object relatedness'. A 'not' is the ground of the 'is'. The *Nicht* stands against as 'resistance', it is a 'preliminary

and constant drawing together into unity'.[253] It is an opening, a 'counter-
standing', a 'primal concept' of an understanding-of-being that projects an
anticipatory *look* 'contrary to the haphazard'[254] a unity amid which beings will
be disclosed.

With respect to the preceding limitation of finite thinking to the task of projecting the horizon for the disclosure of things themselves, Heidegger asks,
seeming to notice the appearance of a reversal of the Copernican Revolution,
'Has the servant not changed into the master?'[255] Yet, he answers his question
with another question: 'Are its mastery and governing, on the contrary, as a
letting-stand-against of these "rules of unity," fundamentally a serving?' The
problem is that of bringing 'mastery', or, the unity of the *a priori* temporal horizon into accord with the meaning of a category as a formal indication of
intuition. Echoing his indication of a pure *sensible* reason, Heidegger contends
that understanding must be '. . . the supreme faculty – in finitude . . .', that it
submits to 'conditions of possibility' to 'remain master of empirical intuition'.[256]
Yet, how does understanding submit? The finite creature, in turning its
attention towards, projects a horizon for a standing against that is a possible
belonging-together. There is a grasp in advance, an understanding, 'through
the horizon of time'. Heidegger writes:

> The unifying unity of pure understanding which grasps in advance, there-
> fore, must itself already have been united previously with pure intuition as
> well.[257]

Openness is the meaning of a thinking which encounters beings. Pure understanding is dependent upon this givenness of a myriad intuition. Finite knowing,
as it is dependent upon the 'freedom' of the thing, projects a unity, as unifying,
before itself in anticipation of a being. Awaiting opens up this possibility for an
encounter. Finite pure understanding indicates, specifically in the context of
Kant, self-consciousness as 'transcendental apperception', a representing unity,
an 'original holding of unity before itself'. The 'I think' indicates an always possible 'consciousness' of 'unity' in general, or pure self-'consciousness'. It says
'I am able', 'I can', and indicates the possibility of an 'essentially free comporting'.[258] The meaning of self-consciousness remains obscure however in terms of
the question of its 'unity'. Heidegger points out Kant's ambiguous statement
that pure understanding 'presupposes a synthesis, however, or includes one'.[259]
It is a conspicuous 'disjunctive phrase' which indicates a 'breach' in the Kantian
text, one which renders the latter susceptible to a destruction. Specifying the
problem succinctly, Heidegger writes:

> Characteristically, Kant wavers here in the unequivocal determination of the
> structural relationship of the unity to the unifying synthesis.[260]

Heidegger pursues this clue and indication which serves as a guiding premonition for his interpretation of Kant's own statements that synthesis is the 'mere result of imagination', or that it is 'formative *a priori*'. Heidegger points out another statement which clearly exposes Kant's susceptibility to *destruktion*:

> Thus the principle of the necessary unity of the pure (productive) synthesis of the power of imagination prior to apperception, is the ground for the possibility of all knowledge, especially of experience. (A118)

In this quotation, excised from the B Edition, pure thinking has before itself, in the act of its unifying unity, a regulating unity of pure imagination and pure synthesis. A representing unity is a unifying of that which is given in advance in pure intuition. For a unity to be possible, there must thus be an *a priori* relationship between imagination and time, and imagination and apperception:

> . . . in this way is [the pure power of imagination] unveiled as the mediator between Transcendental Apperception and time.[261]

Heidegger expresses a distinct reversal of the disjunctive phrase: 'It is only understanding to the extent that it "presupposes or includes" a pure power of imagination.'[262] In this explication of the disjunctive phrase, understanding is exposed as regulated, thrown amid a horizon of time. Indeed, understanding is itself a creature of imagination, and hence, original time. Heidegger quotes Kant:

> The unity of apperception in relation to the synthesis of the power of imagination is the understanding; and this same unity, with reference to the transcendental synthesis of the power of imagination, is the pure understanding. (A119)

The dependence of understanding upon imagination tempts the question: is pure, productive imagination the originator of finite knowing? Heidegger answers in the affirmative and gives an example. He quotes Kant to the effect that in factic experience, the encountered is 'found "scattered and individually"' (A120). Yet, if we are to raise any question of 'connectionedness', a question seemingly implied in the description of scattered and individualized, we must already have a sense, understanding of 'connection', 'unity'. To pre-present connection in advance, however, means: first of all to form something like relation in general by [representing], manifesting it. However, this power, for Heidegger, which first and foremost 'forms' relations – is the pure power of *a priori* imagination.[263]

This pre-presentation, this playspace of connection and connections *in concreto*, is time; Heidegger calls it a 'normative unification', a premonition and projection of 'unity', which, if we consider the conclusions of the *Phenomenologi-cal Interpretation*, discloses transcendental or pure productive imagination as apperception. Heidegger accentuates Kant's comment that it is (A123) 'strange' that affinity, a pervasive unity which precedes the explicit 'unity' of the Kantian 'concept', is given by imagination. All connecting relies on a previous represen-tation of 'unity'. Imagination forms the affinity of letting this be-ing be encountered while pure apperception (the pure I) lets be that which stands against. This 'object-relatedness' occurs in the envisaging of a phenomenology that brings temporality and apperception together in the same intentional field.

Transcendental apperception is dependent upon a 'unity which forms itself in the unifying';[264] it depends upon imagination for this pure *look* of unity. The category contains an original 'unity' of time and thought, pure imagination is the 'middle', which as a synthetic hybrid, or as an original power, has a little of each of the extremes, and is able to bring them together, to disclose their existent(ial) 'unity'. Heidegger quotes Kant (A124) who states that transcen-dental imagination is a

> fundamental faculty of the human soul, which serves as the basis for all knowl-edge *a priori*. By means of this, we bring the manifold of intuition {into connection} on the one hand, and we bring {this} into connection with the condition of the necessary unity of pure apperception on the other.[265]

As we enact the de-severing, as we move closer to the root, this original 'unity' of *being-there*, we can no longer grasp, if we ever did, pure intuition/pure imagi-nation/pure apperception as a mere juxtaposition between faculties. Pure synthesis points to a 'unity' of pure knowledge, to an intentionality of existence. An understanding-of-being, in Heidegger's re-reading of Kant, is grounded in a pure imagination which is 'relative to time'. It is, in this way, a pure *sensible* reason.

The persistent question of authority, or perhaps jurisdiction of the concept becomes the question of the original limits of real predication: to what extent can content (reality [**Realitat**]), that is represented in pure concepts, be a determination, description of that which stands against finite knowledge? For Heidegger, an original 'unity' among differing aspects and comportments with the being itself is rooted in the category. The latter is a 'hinge' (**Fugue**), 'build-ing block' of transcendence; its 'meaning' emerges in its 'performance', in the 'act' of its understanding-of-being. Heidegger accuses Kant of missing the point of the Deduction with his 'juridical' metaphor: such a metaphor emphasizes the question of validity to be determined by the logical functions of judgement. He claims that Kant gets lost in the 'external form of the deduction', just as he seems to get lost in the labyrinth of his own architectonic.

The ground for the inner possibility

Heidegger describes the 'inner possibility of ontological knowledge', as the 'structure of transcendence', which he describes as an opening which is held together by the pure power of imagination. The emphasis, as we find in Kant upon the 'logical-rational way of posing the question in metaphysics' obscures the 'relations' between the pure power of imagination, pure understanding and pure intuition. This question of the inner possibility is disclosed as the unifying operation of pure imagination as a 'self-forming of transcendence and its horizons', setting free the 'essential ground'.[266] This setting free will take place via an examination of the schematism as the source of an indigenous conceptuality of temporal existence.

The schematism as a possible origin of conceptuality is the place for asking 'the fundamental question regarding the transcendence of the finite creature'.[267] This comportment of turning-towards makes possible, in advance, an *a priori* encounter. Moreover, the horizon of an encounter 'must itself have the character of an offering',[268] a pure look, a playspace of openness, or, a letting-stand-against intuitively which offers 'objectivity as such'. With an original groundedness of pure concepts in sense, a finite creature forms its own look of the horizon of 'preliminary turning-toward'. This 'look' is made perceivable for itself in its receiving of the offering, but a receptivity which maintains, with pure intuition, a capacity of projection (spontaneity) via the pure power of imagination. Schematism projects this *look of the horizon*, it '"creates" this horizon as a free turning toward'.[269] It opens up a comportment and gives a schema, a sketch. This double forming of finite transcendence, as an intentional opening and a look, of being and logos, makes 'visible' transcendence, and makes understandable what Heidegger designates as the 'look-character'. Schematism is thus, as we have seen, a pure making-sensible. He writes:

> The pure power of imagination gives schema-forming in advance the look ('image') of the horizon of transcendence.[270]

Pure sensibility takes in stride in advance a look which does not, as a finite knowing, offer the being. Making-sensible is a making-intuitable, to distil or create a 'look from something'. Such a procedure recalls Kant's attribution to Leibniz of direct access to sense in abstractive acts of imagination upon and amid sensibility. It is, in this way, open to the phenomenon as it gathers a look. Heidegger gives us the example of a 'death mask', which is likeness-taking and offers an 'immediate contemplation of a likeness'.

On the other hand, there is a 'look in general' with respect to the making sensible of that *one* which 'applies to many'. A concept points to the indications of unity in the sensible concept, the looks are the *eidos*/idea of the concept.

But, the concept in the sense of a 'look in general', as *hen*, a one, as the act of unifying, is 'not capable of having a likeness taken'.[271] Heidegger gives the example of a house – What is the relation between this-here house and house as such? 'It shows us "only" the "as . . ." in terms of which a house can appear.'[272] He writes,

> If the concept in general is that which is in service to the rule, then conceptual representing means the giving of the rule for the possible attainment of a look in advance in the manner of its regulation.[273]

The concept is not given in a likeness, but it acts through its double, a rule regulating the representing of a *look*. This is not the formal concept of the severed stems but the unity which has unifying as its most peculiar content. Heidegger contends, 'Only as a regulative unity is the conceptual unity what and must be as unifying.'[274] Remembering *ens imaginarium* as Kant's 'regulative' no-things in the First *Critique*, we can detect the *a priori* meaning of a transcendental imagination which makes manifest pure concepts. Making-sensible is neither the simple look, nor the isolated concept; but is the regulating event of making-sensible, disclosed in the 'how'. Heidegger describes this event,

> The representing of the 'how' is the free 'imaging' ['Bilden']of a making-sensible . . . not bound to a determinate something at hand.[275]

A transcendental schema, which 'in a true sense is an image of the concept'[276] makes manifest a rule and is articulated amid possible schema images. A schema as making manifest leads forth (**producere**) a rule into a 'sphere of possible intuitability.'[277] Intuition is a 'procuring of characteristic images'.[278] Such a procurement is a rough sketch of finite knowing without real predicates, a premonition of the sense of being already there, a comportment and a projection of pre-understanding. Or, to put it like Kant, an 'art concealed at the depths of our soul' (A141, B180), or, the 'innermost construction of transcendence'.[279] Time as pure intuition and pure image, is a schema-image and form of and formal intuition. The schema has a 'character of its own' and regulates [concepts] as the pure unities 'internally in time'.[280] Time is a unique 'object': it is intuition, the possibility of having a special kind of *look*, a pure image, a monogram that anticipates the variety of pure concepts. Pure imagination, in this way, projects transcendence, gives an *a priori* 'unity' made sensible in an intentional opening, amid that horizon of standing-against (object), this taking in stride (receptivity) of finite transcendence.

Kant postponed any in-depth analysis of schematism, but merely arranged the twelve schemata, as Heidegger describes, 'according to the order of the categories and in connection with them'.[281] The relation for Kant between concepts and intuition remains that of a subsumption – placing 'objects "under" concepts'.

Heidegger, however, detecting another breach for his *destruktion*, seeks to appropriate and subvert the lexicon of subsumption by asking after the meaning of an *ontological subsumption*, through which an original conceptuality would emerge. An ontical subsumption is a 'bringing' of a particular under a universal. He challenges the logic of this system of words. An ontological subsumption would point to the matters themselves, to the singularity of the projection of being upon temporality in the midst of the self-expression of finite existence. He asks after the 'proper' means of expression for the singular, after the 'destruktion' of the universal, the death of God.[282] It is, Heidegger insists, through the Schematism, the imagination, that any 'rule of unity' is made manifest, that this disclosure amid one's lived existence '"forms" the unity represented in the notion into the essential elements of pure, discernable objectivity.'[283] It is in schematism, Heidegger contends, where concepts arise: In the transcendental schematism the categories are formed first of all as categories. If these are the true 'primal concepts', however, then schematism is the original and authentic concept-formation as such.[284]

   With this genealogy, Heidegger announces: 'The question concerning the inner possibility of original conceptuality as such has burst open.'[285] Yet, 'what' meaning can 'subsumption' have any longer if schematism is the place of 'origin'? An original conceptuality stands outside the limits of ontic genus/species classification, of real and ideal predications. For Heidegger, the schematism is the decisive stage in the ground-laying.[286] The *destruktion*, the deseverance of the stems into the jointure of trunk and root, finds an indication, a preliminary sketch in the notion of finite expression, philosophical *logos* that gives voice to phenomena, untainted by a 'mythology of consciousness'. It will thus be in expression, and not in subsumption, that we will find the emergence of an indigenous conceptuality.

The full essence of ontological knowledge

Since from the beginning we have been seeking to deconstruct the starting point in severance, and since a ground in pure imagination has been uncovered in the deconstruction, Heidegger claims that it is necessary to 'take possession of the ground' which has been 'won as such, i.e., with a view to its possible cultivation'.[287] He writes of a seizure of territory (*topos*), a clearing and cultivating of the seeds, roots, stems of philosophy/thought. He seeks to place the provisional 'stages' into their unity, not any longer as step by step, but, with reference to Dilthey's holistic *a priori*, as an 'ontological unity', a 'unity' of being-there, existence. Access to this 'ground' is indicated, by Heidegger, in the 'highest principle of all synthetic judgments' which exhibits the 'essence of transcendence'.[288] As Heidegger quotes kant: 'The conditions for the possibility of experience . . . are at the same time the conditions for the possibility of the objects of experience.'[289] Heidegger, in his fashion, focuses on the phrase 'at the same time', suggesting that it reveals an 'essential unity of the full structure of transcendence', and in a reference to *Being and Time*, remarks that the living unity of finite knowledge, as transcending, going out to, is 'ecstasis', a 'constant standing out from'.[290] Heidegger again quotes Kant[291]: 'It is but one quintessence in which all representations are contained, namely, inner sense and its a priori form, time.'

This *standing out from* forms and holds open before itself a horizon. Transcendence is thus an ecstatic-horizonal opening for the disclosure of entities amid one's own horizons of concern. It does not only concern itself with objects at hand, but with the prerequisite opening up of an intentional relationship itself as the condition for the possibility of experiencing any 'thing'. Transcendence is in this way no-thing, a 'terminus of the preliminary turning toward . . .' which holds open a horizon, and is '. . . unthematic, but must nevertheless be regularly in view'.[292] Transcendence projects the horizon of 'original truth', of an 'unveiledness of Being and the openness (**Offenbarkeit**) of beings'.[293]

In the wake of his *destruktion* of 'ontology', Heidegger announces, 'With the transformation of Metaphysica Generalis [grasping itself in its finitude] . . . the ground upon which traditional metaphysics is built is shaken, and for this reason the proper edifice of Metaphysica Specialis begins to totter.'[294] The question of the nothing, of the radical temporality of thought, shakes the foundations of tradition criterias of application in that the very elements of such analysis is put out of play, though not eliminated.

Chapter 8

The Articulation of Finite Knowing

Thus the whole of philosophy is like a tree: the roots are metaphysics, the trunk is physics, and the branches that issue from the trunk are all the other sciences . . .[295]

The fruit of the preceding *destruktion* is an uncovering of the imagination as the 'formative center of ontological knowledge'. As a 'surrogate' for ecstatic temporality, the imagination projects an original *topos* of expression for the self-interpretation of existence, Dasein. Heidegger insists that his *destruktion* of the 'stems' to the 'root' must transcend the architectonic limits of the three *Critiques*, and must move towards the question of the retrieval of 'transcendental philosophy' as such. In this way, we will seek in the following, to understand Heidegger's re-writing of Transcendental Philosophy.

Moreover, just as Heidegger's deconstruction of the First *Critique* made a reversal in the order of origination from the derived elements, of sensibility and concept, to the schematism, his projection of the originality of pure imagination transforms the significance and intelligibility of the Second and Third *Critiques*. His destructive appropriation is not intent upon orchestrating a system of self-consciousness from the Kantian fragments in the manner of the German Idealists. Nor does he simply remain with the First *Critique* to the detriment of the other aspects of philosophy in the manner of logical positivism. On the contrary, Heidegger asserts the radical primacy of temporal existence through a radical interpretation of the primacy of the practical in Kant's own philosophy. With his indication of Care, however, and the 'ground' of the latter upon ecstatic freedom, Heidegger places the 'authority' and the alleged autonomy of reason into question within the horizons of his temporal problematic.

The indication of the 'common root' has invoked a metaphor to describe the morphology of finite thought.[296] According to the metaphor, pure imagination or original temporality is the *common root* of the gestures of the stems, of sense and concept. In his *Phenomenological Interpretation*, for instance, Heidegger uncovers the 'acts' of transcendental imagination already 'there' in the First *Critique* in the Transcendental Aesthetic in *syndosis*. Syndosis, as we have seen, points to the transcendental synthesis occurring 'prior' to apperception, which is the implication of the *Preliminary Remark* in the A Edition, that there is an original apprehension and a 'coherent connection', generated by synthesis prior to any recognition by the concept. It is a collocation, not yet a determinate connection of in the sense of strict knowledge, but a 'first rudimentary knowledge' needing only to be analysed, a 'makeshift' knowing, one which Kant disparagingly portrays as the loose schemas of a painter or physiognomist. The priority of pure imagination casts light upon a pre-theoretical and pre-practical intimacy amidst pure intuition, prior to the concerns of a mere 'aesthetics', considerations which only became prominent, as Heidegger argues, after the corruption by Baumgarten.

Continuing our readings of the Kantbook, we will thus travel along with Heidegger upon a pathway that seeks to articulate the essence of a Transcendental Philosophy that has revealed to itself the pure imagination as its ownmost ground. Without explicitly re-stating the question of the meaning of interpretation, once again, such a reading of the First *Critique*, not only discloses one of its hermeneutical potentialities as a text, but also allows us to glimpse that which seemed so daunting to Kant and to the Neo-Kantian critics of Heidegger. All in all, that which we are given as Transcendental Philosophy becomes a phenomenology of temporal existence in which pure imagination is acknowledged as a primary condition for experience and thought.

The bifurcated ground of finite knowing

The disclosure of the ground in imagination, for Heidegger, answers the question of essential 'unity' of finite, ontological knowledge. This 'unity' as transcendence is the holding-open of the horizon, disclosed via the pure schema, which are pure synthetic constructions of pure, productive imagination. Heidegger re-iterates: 'As original, pure synthesis, it forms the essential unity of pure intuition (time) and pure thinking (apperception).'[297] Transcendental imagination is the 'indispensible function of the soul' which builds an *a priori* opening for our access to 'matters', amidst this 'riddle of fallenness, the building site of care and temporality'.[298]

To gain a deeper grasp of the possibilities of the 'Kantian' imagination, Heidegger takes us on a short 'detour' through Kant's lectures on Anthropology. As a 'detour', he seeks to disclose potentialities in the Kantian text which are rarely emphasized as they do not fit the plaster saint idol of the 'real' Kant. In his lectures, Kant, in a way that became typical after the *Prolegomena*, restricts imagination to the realm of intuition, to sensibility, deemed, in the manner of Plato, as a '"lower" faculty of knowledge'.[299] Heidegger accentuates however Kant's notion of an intuitional imagination, which is capable of an 'intuition "even without the presence of the object."'[300] For there to be an image, the being need not be even 'presenting'. Imagination, for Kant, though a lower faculty, has a 'peculiar non-connnectedness to the being', 'without strings'. And, perhaps this is why Kant disparages this faculty, just as Plato excluded the poets from his *polis*. In its disconnection from the object, the imagination 'lies too much'. Heidegger, turning this humiliation of imagination upside down, insists that it is the pure imagination which 'gives itself such looks',[301] gives an image that it is free to form without the presence of an object.

Pure imagination, as a 'formative power', is in this way simultaneously a 'forming' which takes things in stride (is receptive) and one which creates (is spontaneous).[302] It is here that Heidegger detects, in the act of the imagination, the possibility of an *a priori* horizon which opens without the presence of the object, and is thus that which projects the horizons for the entry of objects into 'world'. Contrary to Kant's wishes, and indeed, arguably turning these on their head, Heidegger has disclosed an iridescent imagination as the 'middle' betwixt the two bifurcated stems. It has this peculiar power to be the root of the stems in that it contains within itself the elements of spontaneity and receptivity. Imagination, as with the schematism, it is the 'third thing', 'homogenous to both', which projects a horizon of significance. Heidegger explains the pre-theoretical character of this thought in the middle,

> 'Imagining' then, refers to all representing in the broadest sense which is not in accordance with perception: conceiving of something, concocting something, devising of something, wondering, having an inspiration and the like.[303]

Imagination represents present intuition, *subjectio sub aspectum*, but, in that it has this bi-character of receptivity *and* spontaneity, it can also become memory and anticipation, *exhibitio derivativa* and *exhibitio anticipatia*. Only when it freely composes, *exhibitio originaria*, is it productive, not in the sense of divine creation, but as the projection of 'forms', of this 'look' of a 'thing', as an *a priori* horizon, as a turning-toward. Heidegger, revaluing Kant's expulsion of the imagination, discloses 'imagining' as the projection of an intentional opening of finite transcendence, the no-thing, without, outside real or ontic production, perception and predication. He describes this event of opening,

> The imagination forms the look of the horizon of objectivity as such in advance, before the experience of the being. This look-forming (Anblick-bilden) in the pure image (Bilde) of time, however, is not just prior to this or that experience of the being, but rather always is in advance, prior to any possible experience.[304]

The imagination, as '. . . originally pictorial in the pure image of time',[305] projects the schematic horizons for the emergence of beings in the world and their own self-expression. Yet, as original temporality, the imagination gives rise to time, but not in the sense of Plato, in his *Timaeus*, who referred to time as the moving image of eternity. For Heidegger, temporality, as a projection, as an 'act', 'forms' an opening of transcendence, and, is not thus an image at all.

Temporality is the 'origin' of the image of the time on the clock, which is a generic emblem and artifact of the horizons of *our* finite existence. Yet, Heidegger asks, 'How does the going back to the origin occur?'[306] and answers by setting a task, '. . . to deduce the direction of the going back required by the dimension of origin itself and to do so from the already-laid ground itself.'[307] The retrieval of the origin will be oriented by the *topos* of the 'matters themselves', disclosed amid our pre-understanding, and will be understood as setting-free original temporality, or a return to that which is one's ownmost possibility. Indeed, we have found that the meaning of the 'origin' will be disclosed in the self-expression of finite existence from within the limit situations of original temporality. It is in the 'nowhere' and the ecstatic silence of a return to the self, as we will see in Chapter 13, where worlds are born, where a makeshift world of binding commitments is resolved, projected as 'there' for the time being.

Transcendental imagination as the third basic faculty

In order for transcendental imagination to be regarded as a faculty, Heidegger considers the phenomenological evidence for such a designation. A faculty, he contends, is a capacity, a 'making possible of the essential structure of ontological transcendence'. It is not to be conceived, one of the points where Henrich agrees with Heidegger, as a 'basic power' in a soul. Such a notion betrays an ontic character, and not that of an *a priori*, ontological making-possible. Amid the re-configuration of finite knowledge, with the deconstruction of the unworlded stems of the theoretical back to an original clearing (**Lichtung**) of 'world', the 'faculties' are re-designated as indications of lived existence. Kant had stated that there are two stems or sources of 'mind', with perhaps a 'common root', and Heidegger points out that this binary division corresponds to the division into Aesthetic and Logic. Yet, in such a predicament, Heidegger laments, 'The transcendental power of imagination is homeless.'[308] Pure imagination, he argues, is not articulated in the transcendental aesthetic 'where it properly belongs' (although, I would point out that it is thematized in Kant's criticisms of the notion of indictive imagination as a 'framing out . . .' of relations in Leibniz and Wolff). Indeed, Heidegger implies that there is an obvious confusion in the First *Critique*, most conspicuously manifest in Kant's treatment of imagination in both editions of the Analytic. Not even asking the usual questions, one need only ponder Kant's inability to clearly *place* the imagination, which Heidegger casts into relief.

In a significant way, and echoing Sallis' *Spacings*,[309] Kant's problem, his confusion, is exposed in his deployment of the metaphors of 'sources of-' and 'stems of' knowledge as beams in the scaffolding for his architectonic. Kant's reticence with respect to the implied root of the stems, or, of imagination as a source for knowledge, is broken open in the course of the A Edition and in unaltered sections of B, such as the Schematism, where he clearly continues to designate imagination as the *third* – as the root for 'the stems'. Kant, in other places, describes concept and intuition as 'sources of knowledge' and designates, via Aristotle, imagination as a mediator, as a supervenient link, a bridge. Is it accidental or irrelevant that Kant uses this image to characterize sensibility and understanding, or is it instead used just to indicate that they grow from a 'common root?'[310] Does this conflict in metaphors, as Sallis argues, expose the almost undetectable links in texts which claim only to be governed by reason? Can we deploy the metaphor as a formal indication and see where it leads us in a phenomenological interpretation? It would seem that we do have this freedom, as long as we are transparent with respect to our hermeneutic method. Even disregarding the significance of metaphor, however, the problem of the homelessness of imagination, not only deserves scrutiny, but also suggests the possibility of imagination as an *a priori* 'faculty' of the soul.

Heidegger's *destrucktion* finds its orientation, as we have seen, not in the traditional placement of imagination between, as a mediator of the 'unworlded' stems, but instead, as an 'original unifying'. This 'unity' is not that of a concept, but of a transcendental power of imagination, as a formative centre of knowledge that coordinates its equiprimordial moments. This interpretation of imagination as the root of both stems, and as Temporality, as we suspected all along, intimates a crisis amid traditional grounding in the 'unity' of 'reason', of logic. The 'stems', in another context referred to as 'sciences',[311] have their ground, rooting, root-being (**wurzelsein**), in the transcendental 'unity' of experience. The pure power of imagination is an excession beyond the limits of logical predication into self-expressive existence.

Heidegger contends, pre-empting suggestions that his interpretation would imply that knowledge is merely imaginary, that the stems are rooted structurally in the transcendental imagination. Products of 'mere' empirical imagination are ontic, and presuppose an *a priori* or ontological imagination, which in phenomenological terms, is a playspace of disclosure. At the same time, cutting off his translation of the imagination into the field of his ontological difference, Heidegger calls into question the indication of the designation of a 'power of imagination', as susceptible to becoming just another being, thing or faculty. To avert this threat of reification, Heidegger excavates transcendental power of imagination 'into more original "possibilities" so that by itself the designation "power of imagination" becomes inadequate . . .'[312] Yet, as with other artefacts of Kant's text, it can still be interpreted as a formal indication, but the term will maintain the traces of disputed role in 'transcendental philosophy'. In consequence of his destruction and of the necessity for novel expressions or grammars of existence, Heidegger writes that, in light of his temporal problematic, there can be no *absolute* explanatory basis, but only a marked strangeness, ourselves to ourselves, as 'the most unknown and the most actual . . .'[313]

We need, therefore, to place into brackets all that which is familiar and attempt a re-articulation of Transcendental Philosophy.

Pure imagination and the stems: Pure intuition and pure thinking

Heidegger suggests that Kant must have 'understood little' of pure intuition. Yet, he admits that Kant intimates this 'in advance' character of the conditions, or, horizons, of possible experience, not as beings, but as *ens imaginarium*, 'species of the nothing' that are a 'formative giving'.[314] In this light, and as another reading of Kant's lectures on anthropology, pure intuition as a 'look', is pure, unthematic and unobjective, a pre-theoretical/practical disclosure of the horizons of existence. Kant, remaining it seems with the familiarity of the dogmatic 'object', can be faulted only for his failure to pursue a discovery of which he may not have been aware – or wished to avoid.

The pure intuitions are original, for Heidegger, in that they allow 'things' to spring forth, 'they pro-pose (**vorstellen**) the *look* of space and time in advance as totalities which are themselves manifold', in an original exhibition, *exhibitio originaria*. Pure intuition is in this way an *a priori* projection which allows 'something' to spring forth. As it is original, Heidegger argues, it is essentially the 'same' as the pure power of imagination which 'formatively gives looks (images) from out of itself'.[315] Pure intuition, as we have seen, is a whole, whose parts are limitations of itself, and thus, it is 'originally unifying'. There is thus a 'unity' in pure intuition which must be 'caught sight of' in a seeing together (*synopsis*), amid a comportment, distinct from, but essentially related to, and grounded upon, the synthesis of the power of imagination. Such a 'unity' is not that of the concept, with its seat in the Kantian 'understanding', but is 'caught sight of in a glance of an eye, in advance in that which Heidegger indicates as an image-giving imagining [**im Bild-gebenden Einbilden**]',[316] a 'moment of vision' amidst the horizons of an original or ecstatic temporality. Through a 'syndosis' of imagination, the sister both of 'synopsis' and 'synthesis', pure intuition pre-forms its own horizon of perceivability as a being-perceived.

Heidegger, as we have seen, agrees that this reduction of thought to imagination seems to be absurd. Yet, as finite intuition is 'taking in stride of what gives itself',[317] this absurdity of origination 'falls' when we consider that imagination is not merely sensible in the brute sense, or, even in the *a priori* sense of pure intuition. In its own *a priori* pure significance, transcendental imagination is the 'ground' for all the 'faculties', as the formative centre of ontological knowledge. As this status was unearthed in the deconstruction, the origination seems less untenable 'even if importance can no longer be attached to the order of precedence of sensibility and understanding'.[318] Pure imagination and pure understanding are species of representing. As such, pure thinking exposes itself as 'dependent' upon that which gives itself, a 'dependency upon intuition'. This dependency as a state of being discloses the understanding as a being which is essentially caught in the horizon of imagination. Heidegger suggests that it is only logic, 'which does not need to deal with the power of imagination', which gives 'reason' alibis of independence from an infection of temporality, of imagination. Indeed, Kant moves towards transforming the character of Transcendental Philosophy. Yet, Heidegger charges that Kant remains entangled within that severance of a 'formal logic' and that it is Kant's 'derivation' of 'concepts' from logical functions of judgement, and his subsumptive procedure of understanding, which incarcerates imagination within the regime of a pure, independent thinking. Heidegger counsels, 'The analysis must indeed depart from this independence of thinking if the origin of thinking in the power of imagination is to be shown.'[319] Questioning the ground and meaning of thought itself, Heidegger contends:

> Only by beginning with the original essence of understanding, and in no way with a 'logic' that slights this essence, can a decision be made concerning its possible origin.[320]

Everything is thrown up into the air. Thought projects, proposes (Kant), and in the re-writing, it 'holds in advance' a finite horizon of possibilities which are *ens imaginarium*, species of the nothing. In this light, Heidegger intimates the ecstatic character of the self, 'In such a proposing [vorstellenden], a self-orienting towards . . ., the "self" in this orienting-toward . . . is, as it were, taken outside.'[321] Taken outside, the 'I am', Sum becomes 'to be', Esse, it exists, and in that it exists, is ecstatic, projecting a horizon of possibilities to which it transcends amid its 'worlding'. 'The "I" "goes with" in the pure self-orienting.'[322] The 'I' is never that detached emptiness of formal logic, but is intentionally related as its own opening of possibility: I think this, that . . . 'The I is the "vehicle" of the categories' in this self-orienting, pre-forming of a horizon of unity 'from out of itself'.[323] The spontaneous projections of horizons is the operation, for Heidegger, of schematism, there is an intimacy of thought with schematism, with imagination, and hence with temporality; it 'works with them'.

In the wake of this intimacy, and rejecting the rootlessness of severance, Heidegger takes us to what for him has become an inescapable conclusion: the *understanding* itself, in its 'original Being', *is* pure schematism, transcendental imagination. As with the *Basic Problems*, an understanding-of-being, in this sense, is an 'unthematic bringing-itself before – us – of that which has been proposed'.[324] Heidegger's re-defining of thought, 'in the sense of a free-forming, projecting, although not arbitrarily, "conceiving" [**Sichdenkens**] of something . . .' implies that pure thinking must be the same as pure imagining, temporal thinking – and is not to be confused with 'logic' traditionally conceived. Indeed, Heidegger could suggest, even logic with its acts such of inference, could be conceived as an imaginative activity, with its own rules. In light of this re-valuation of imagination, Heidegger contends that, contrary to Kant's denial of Ideals to the imagination, pure imagination could possibly 'form', if we are to remain within this conceptuality, 'ideals' for the future. In temporalist expression, however, ideals would acquire the sense of projections of anticipatory resoluteness, and thus would disclose an 'event' of original 'unity', and, intimacy, of self-interpretation. In this radical transfiguration of knowing, transcendental imagination becomes an original 'unity' of receptivity and spontaneity, issuing forth conceptualities of existence, topologies of expression amidst a self-interpretation of Dasein. Pure receptivity is open to a spontaneity of that which gives itself from itself. Only a 'logical', that is, empty, conception of reason may be merely spontaneous, a mere playing with concepts.

As suggested earlier in our exploration of the temporal horizon of *a priori* knowing, the intuitive character of pure thinking necessitates a constant unifying in advance of the 'haphazard'. Yet, even this projection of rule discloses another face, contrary to some commentators such as Weatherston, that of its own receptivity and finitude in the wake of its own horizonal and topographical character. Thinking is an index of finitude, and as imagination, it has a 'unity' of receptivity and spontaneity. Pure thinking as pure sensible reason, is thinking in the limit horizon of finitude. 'It' projects horizons for a possibility of this self, this 'I' that exists. One is free for a 'self-given' dedication of oneself to 'self-given necessity'.[325]

Pure imagination and practical reason

Towards the end of his Kantbook, Heidegger sets out a concise interpretation of 'what is essential' regarding the *Critique of Practical Reason*. And, as we will see, it is the projection of *ens imaginarium, a priori* horizons of discernability, which will provide the traces by which he can excavate imagination in Kant's doctrine of 'true morality'. He begins with a link between respect and the moral law:

> Respect as such is respect for the moral law. It does not serve [as a basis] for the judgment of actions, and it does not first appear after the ethical fact to be something like the manner in which we take a position with respect to the consummated act. On the contrary, respect for the law first constitutes the possibility for action.[326]

Respect must come *first*, it must await the coming of laws. As anticipatory, and spontaneous, as a 'first', respect is on the one hand isomorphic to the indigenous, pre-theoretical and pre-practical anticipation of the 'matters themselves', being open-for. On the other hand, respect exhibits a receptivity which projects the horizon of limit, a 'toward which according to which' in which Limit, *peras* (πέρας) is 'accessible to us' amid an open encounter of 'world'.

The self is 'I', 'self-consciousness'; but the meaning of the self is determined via finite existence, the be-ing of this self, which *manifests* a world, and in which its manifestness 'co-determines the being of the self'.[327] This image, or manifestness of the self, is the understanding of the being of the practical, moral self ('authentic self'/'essence of man') or, the person. In light of this 'unity' of the spontaneous and receptive aspects of the understanding-of-being, Heidegger contends that the being of personality must find its origin in the transcendental imagination, despite its suppression in the Typic of Practical Reason. Personality, as we saw in the context of Husserl and Scheler, is the idea of and respect for the moral law. 'Respect is "susceptibility"' to the moral law, and as self-determination, it is a way of 'Being self-conscious.'[328] That which lies beneath the numinous ideas of personality, respect and the moral law, however, is, for Heidegger, the ecstatic be-ing of the person, whose free act is that of a projection amid horizons of self-limitation. In this context, Heidegger digs deeper for a 'pure feeling' of respect beneath the sense of feeling with a merely sensible character in the Kantian designation. He translates this *pure feeling* into his own indication of disposition (**Befindlichkeit**), one which can be a way or mode of 'Being self-conscious'. In this context, he invokes the 'temporal category', the existential, as for instance, that of being-towards-death in which anxiety, a 'feeling', 'state of the soul', discloses that which is 'there'. All feelings, Heidegger writes, are beyond the feeling itself, as a way of being self-conscious: feeling for . . . and a self-feeling, disposition. Self-feeling is a feeling-for. Respect, as this feeling of the moral self, is a pure feeling. In a feeling of respect, the self discloses itself to itself. Heidegger writes: '. . . respect before the law – this determinate way of making finitude manifest as the determinative ground for being – is in itself a making manifest of myself as acting self.' This self-disclosure occurs amid the horizon of personality in distinction from things. Respect entails an acceptation of limit for oneself. Yet, this limit (which is not of logic, but of existence) is at the same time an openness to the possibilities of one's self, of a finitude 'grounded' upon an originary *topos* of freedom.

Heidegger admits that Kant conceives this limitation as a 'giving of a law' to the self. But, since this 'law' becomes merely the equipment of a 'practical' reason that has from the outset refused sense and temporality, this subsumptive self-determination thereby nullifies existential freedom, and with reflection, cannot be a 'lawfulness' that has been *given* to oneself amid a free existence. Respect, in Heidegger's re-writing, intimates an ownmost (**eigentlich**) selfhood. 'I' project myself upon my ownmost horizons of possibility, thereby making manifest my own *topos* of limit. The limit is disclosed as the resistance of the 'law'. This 'law' is disclosed, in turn, amid a limit-*situation*, as 'acting Being itself'. The feeling of respect named by Kant as 'practical reason', becomes the 'basic structure of the transcendence of the moral self'. This transcendental 'sense' of feeling discloses pure imagination as pure self-affection, as a temporal disposition. Heidegger writes,

The self-submitting, immediate, surrender-to . . . is pure receptivity; the free, self-affecting of the law, however, is pure spontaneity. In themselves, both are originally one.[329]

For Heidegger, respect occurs as the self-affection and self-disclosure of the being of the self as a finite being. In the original oneness of receptivity and spontaneity, the self projects its ownmost possibility amid the self-disclosure of its own limits. Receptivity and spontaneity are once again disclosed as aspects of a 'unifying' imagination, of an ecstatic temporality where we always already comprehend an originary *ethos* (cf. 'attunement' in Ancient Greek musical theory) of finitude. Heidegger writes,

> only this origin of practical reason in the transcendental power of imagination allows us to understand the extent to which, in respect, the law as much as the acting self is not to be apprehended objectively. Rather, both are manifest precisely in a more original, unobjective and unthematic way as duty and action, and they for the unreflected, acting Being of the self [Selbst-sein].[330]

Kant suppresses imagination and temporality in his Second *Critique* as practical 'reason' indicates a direct determination of reason over the will. This determination, transcending the limits of theoretical reason to the horizons of possible experience, intimates the erasure of imagination and temporality in Kant's doctrine of morality. This morality may not be 'strict knowledge' beyond the theoretical face of 'reason', but it has legislative enforcement orchestrated across the networks of local, nodular power. That which may be expressed as Kant's lapse into 'Spinozism' ensures that his self-determination of the will is not a self-giving of the law by the transcendental imagination, temporality, but an indication of Kant's abiding loyalty to an 'authority of reason' and this *pathos*, island of logic.

Pure imagination and aesthetical reason

In notes appended to the Fifth Edition of the English version of the Kantbook, there is a marginal comment by Heidegger that the Third *Critique* is necessary for a completion of Kant's project 'according to its own criteria'. The *Critique of Judgement* is the 'third' that is conceived as a 'mediation' between the theoretical and the practical. Yet, despite its current reputation as an imaginative work, it could be argued that this work merely consolidates Kant's suppression of imagination, the phenomenon, 'matters themselves'. The problem, for Heidegger, is the absence of a *pure*, productive imagination, an absence that conforms to the B Revisions of the First *Critique*, the *Prolegomena*, the *Groundwork* and the Second *Critique*. This absence completes the erasure of the transcendental imagination.

Heidegger contends in his Notes, moreover, that his own thesis is not contradicted by the Third *Critique*. Indeed, he never made any attempt to hide his relation with the Third *Critique* – his openness is *telling*. But, that certainly does not mean that their respective theses are the same. Heidegger suggests that not every productive imagination is *pure*, or in the present context, that an imagination which submits to understanding is not pure, productive or transcendental imagination, or an 'understanding of being'. For Heidegger, Kant is blind to the truth of imagination, of the artwork, as he is to the lived existence of oneself, an apprehension of the overwhelming being-towards-death and an existential coming to a stand of oneself. For Kant, the work of art, imagination, is only allowed to function as an *exemplar* of the *moral law* in its beautiful and sublime aspects. Through this artwork, we would have a feeling for the 'kingdom of ends', as we agree, amid a *sensus communis*, to what is beautiful and what is ugly. There may be 'no accounting for taste', yet, for Kant, as each person as a 'man like other men' has an ultimately 'identical' feeling for 'beauty, truth and good'. As it is confined within the generic productivity of the Anyone, and is made to serve the reproduction of this Anyone, imagination 'within the limits of reason alone', within the *polis*, cannot account for the incessant insurrections of ecstatic temporality.

As Heidegger asserts in *Basic Problems* and elsewhere, elegance and formal symmetry is not the goal of phenomenology as it must remain attuned to the phenomena, even if monstrous. Finite knowing is thus radically reliant upon 'truth events' amid the overwhelming predicament of existence. In this way, art is not merely to serve to ornament the *mores* of existence, but becomes, for Heidegger a method in and of itself for a disclosure of the truth of existence. That which is essential in art however is the event character of this disclosure of truth. It is in this context that we can question Zizek's contention of Heidegger's alleged blindness to this sublime.[331] One could of course argue that being-towards-death shares a family resemblance with the sublime, and shares with it the event character by which each discloses the being of the self. In each case, there is a radical coming to be of the self. Yet, for Heidegger, the self does not emerge, as it is with Kant, with a severance from the sublime, but, on the contrary, in a resolute return to the abyss. Kant uses a military metaphor of war with the sublime, but a drawing back, a tactical retreat, as a self-reflexive severance from that sublime, as if to suggest that one is more powerful than the sublime, since it is merely my representation, and that it is other than it, can separate itself as it sets upon an autonomous ground. Heidegger, on the contrary, in *Being and time*, for instance, sets forth a differing 'origin' for the finite self. It is not to draw back from the sublime, and retreat to a safe island, but instead, to not only hold oneself out into this no-thing, but also, to *fail to hold up* and submit to this overwhelming situation of finitude. Heidegger emphasizes the unexpectant, the surprise, as the 'event character' of the work of art, that which 'in its light' discloses, for the first time, 'what is there'.

Sallis comments in *Echoes – After Heidegger*[332] that it is art, and not, architectonic control that must determine any sense of the *a priori*. Or, in other words, there need not be an imagination 'within the limits of reason alone'. An artwork erupts amid a fragility of ecstatic temporality, it can neither be addressed in an original way via real predication, nor can it be enslaved to the makeshift 'mores' of an era. Art is the *poiesis* (ποιησις) and *praxis* (πραξις) of temporality. Finite thought is radically temporal, and as a pure *sensible* reason, at once, spontaneous and receptive. An artwork is a 'breach' of that 'normality' of an absorptive being-in-the-world. Such a disclosive disruption uncovers that which is 'there', as an earthquake reveals faulty building sites.

Kant's shrinking back from the abyss

The transcendental power of imagination is the 'unknown', a 'disquieting in what is known' – and it is, for Heidegger, the ground for the finite transcendence of pure knowledge. The references to disquiet, and to the unknown are taken from Kant's Introduction to the First *Critique* to make, a rhetorical, but not necessarily unjustified point. Heidegger claims:

> . . . Kant did not carry through with the more original interpretation of the transcendental power of imagination; indeed, he did not even make the attempt in spite of the clear, initial sketching out of such an analytic which he himself recognized for the first time.[333]

The claim is that in 'shrinking back' from what he had discovered, and burying (although sloppily) in the B Edition, Kant unwittingly threatens the entire foundation of his own project. Yet, a 'double bind' is revealed in that, if he did carry through with such an interpretation of the primal status of imagination, then his project would also be undermined. If Kant does not give imagination the status of a formative centre, of the 'middle' with an intrinsic, structural unity, in the sense of a root of the stems, he cannot give an account of philosophical 'unity'. Yet, he cannot maintain this perspective if he wishes to conserve the priority and authority of his island of 'reason', threatened by the 'flood'. Heidegger, reminiscent of the early pages of Nietzsche's *Beyond Good and Evil*, writes:

> How is the baser faculty of sensibility also to be able to constitute the essence of reason? Does not everything fall into confusion if the lowest takes the place of the highest? What is to happen with the venerable tradition, according to which Ratio and Logos have claimed the central function in the history of metaphysics? Can the primacy of logic fall? Can the architectonic of the laying of the ground of metaphysics in general, the divison into Transcendental

Aesthetic and Logic, still be upheld if what it has for its theme is basically to be the transcendental power of imagination?

And: 'Does not this groundlaying lead us to an abyss?'[334] Heidegger claims Kant was on the verge of a revolution in the sphere of moral philosophy, as 'first philosophy', but was impaired by his sensibilistic interpretation of imagination. Not grasping the productive essence of pure imagination, he could draw no (acknowledged) quarter from its wealth of synthetic power. Kant was drawn increasingly 'under the spell' of a traditional conception of reason, seeking to flee an empiricist ethics of happiness and others like Jacobi and Hamann who rejected the authority of reason and instead embraced imagination and language. Heidegger is however seeking something else besides. In his 'identification of imagination, original time and apperception in *Phenomenological Interpretation*, he discloses a radical temporalization of thought. The question is of a pure *sensible* reason, of a "reason" which exhibits the intrinsic character of receptivity, and of a spontaneity which acts in the service of and within the horizons of this intuition, of "time."' That which is primary in Kant's decision in favour of pure understanding 'as opposed to the pure power of imagination is the desire to preserve the mastery of reason'.[335] For Heidegger, on the contrary, 'to us humans at least', finite knowing cannot transcend the horizons of temporal existence. Imagination is not a supplement, but must, for Heidegger, open up an 'essential horizon'. It is a common root which is originally 'there', before 'stems' of sense and concept, as the temporality of the self in its own singular being-in-the-world. In this way, finite knowing becomes a radical phenomenology of factical existence, lived among the 'things themselves', projecting itself as the no-thing of the *a priori* horizon.

Chapter 9

Transcendental Imagination and Ecstatic Temporality

> ... *time is not just what makes transcendence possible, but that time itself has in itself a horizonal character; that in future, recollected behaviour I always have at the same time a horizon with respect to the present, futurity, and pastness in general; that a transcendental, ontological determination of time is found here, within which something like permanence of the substance is constituted for the first time.*[336]

Across the preceding chapters, the 'identification' of pure imagination and temporality has been explored, but mostly in a *destruktive* and appropriative manner. The sense has been given, to some extent, that Heidegger's hermeneutic of the Kantian text is somehow from the 'outside', as an external criticism, unjust, as a foreign invasion, and thus, violent. However, in light of our examination of the alternative typology of interpretation operative in the temporal analytic of Dasein, it could be argued that Heidegger has endeavoured to explicate possibilities which are 'there' in what I have called the *susceptibility* of the text. These possibilities may have never been articulated, or may have been concealed. Yet, they become disclosed for us in light of the breach of Heidegger's questioning. In this way, it would be misleading to consider his *destruktive* translations of Kantian indications as an unjust and violent reduction of 'Kant' to 'Heidegger'. For what is 'Heidegger', after all? Instead, we will explore the affinities of temporality and imagination, not as substitute terms, but as a disclosure of the family resemblance of the common rootedness of intuition and concept in imagination, and of generic space-time in temporality. With the disclosure of a common *topos*, Kant and Heidegger can be invited to speak to one another.

In the following pages, we will examine the 'translatability' of imagination and temporality, with respect to the resemblance of temporality as pure self-affection (KPM) and ecstatic self-projection (BT et al.). After an exploration of Henrich's *The Unity of Subjectivity* against the background of Heidegger's 'unity of ecstatic temporality', I will explore the latter's 'plurivocity' of expression, in that *being is said in many ways*, with respect to pure imagination and 'original time'. We will juxtapose, in this light, the syntheses of transcendental imagination of the Kantbook and the ecstasies of temporality of *Being and Time* and

Basic Problems. The 'temporality' of the three syntheses of imagination, and the schematism of pure imagination, disclosed as pure self-affection, will be set forth in light of disposition (the 'mood structure', **Befindlichkeit**) of ecstatic temporality and its three horizontal ecstasies, so as to cast into relief the 'translatability' of these 'powers'.

Henrich and the 'unity of subjectivity'

Of the post-War critics of *Kant and the Problem of Metaphysics*, it is Henrich who was the first to pursue the positive significance of Heidegger's *destruktion*, if only to indicate a definite difference between Heidegger and Kant. Henrich begins with a criticism of Heidegger's reading of Kant as transgressing the *rules of philology*, a charge, as we know, was admitted by Heidegger himself. In addition to his criticisms, Henrich asks the further question: what then is the *meaning* of Heidegger's 'interpretation' of Kant? In order to expose the 'strange' character of Heidegger's *destruktion*, he lays out an outline of the equiprimordial moments of Heidegger's existentialist analysis in juxtaposition with the overt temporal significance of the syntheses of the transcendental imagination. He seeks to show that Heidegger's interpretation of Kant *is* Heidegger *in drag*. Yet, as we will see, such a criticism fails to not only take into account the peculiarity of Heidegger's method, but also to consider, in light of this method, whether or not Heidegger's interpretation does in fact articulate the *unsaid* of Kant.

Henrich makes two charges against Heidegger's 'interpretation' of Kant. On the one hand, Heidegger's interpretation is alleged to exhibit a a-historical character. Alluding to a possible meaning of his metaphor, Henrich contends that Kant was not an advocate of the thesis of the 'common root' as conceived in the eighteenth century. He insists that Heidegger remains silent about the history of the controversy with respect to a single 'basic power' (which, against Henrich, is not true with respect to his historical references in *History* and the other works of this period), and thus, his interpretation of Kant neglects the historical context, indicated in Kant's *Anthropology*, which stand in the background of Kant's rejection of the designation of a 'common root'. On the other hand, Henrich castigates the 'violent' character of Heidegger's interpretation. Henrich invokes Kant's 'quid juris?' By what right can Heidegger justify his reading of Transcendental Philosophy as a 'retrieval' of the meaning of being in light of his own radical phenomenology in *Being and Time*? Or, is Heidegger's 'destruction of the history of ontology', to be trusted as good scholarship? Henrich's first charge, that Heidegger neglects the history of the question and fails to consider Kant's *Anthropology* is, as we can already see, clearly inaccurate. An oversight in Henrich's criticism is his failure to acknowledge that Heidegger, in his characterization of the common root as the ecstatic play of the temporality, is already, on his own terms, directly exploring the question of history. Indeed, it is original temporality that lies at the heart of Heidegger's interpretation of

Kant, as Henrich knew very well. It is also clear as early as *History* that Heidegger is not simply re-fighting the eighteenth-century battle between the faculties of reason and imagination, and does not intend such a meaning in his use of the metaphor. He clearly came down on the 'side' of Kant in this dispute, but, perhaps, this was already *his Kant* and not the advocate of the 'authority of reason'.

Moreover, despite his criticism of Heidegger's failure to consider the historical context of Kant's decision, Henrich proceeds to admit that neither Heidegger's notion of a 'fundamental ontology', nor his claim of the transcendental imagination as a 'common root', are to be interpreted in the sense of an eighteenth-century 'basic power'. Henrich writes that meaning Heidegger's claims should be instead interpreted from the perspective of the ecstatic unity of *Being and Time*. In this way, granted that Heidegger is not merely repeating atavisms that he knows Kant has rejected, and that we are well aware of the problematic of interpretation as such, then it is possible that Kant and Heidegger may frequent similar *topoi*, not only in regard to their rejection of an eighteenth century 'basic power', but also in their mutual exploration of the relationship betwixt the 'unity' and the temporality of thought. In this way, the question is shifted from the criteria of Kantian philology to the problem of finite, human transcendence.

It is not at all certain that Henrich succeeds in his ambition to 'refute' Heidegger, and those, like Makkreel,[337] who merely defer to Henrich's 'refutation', may wish to take a closer look. The first charge becomes largely irrelevant as neither thinker can be said to be concerned with an eighteenth-century 'monism of power'. As for the second, Henrich fails to raise the question of whether a schema of equiprimordial moments may not also be a plausible interpretation of Kant, as has been argued skilfully by Sherover and others. It is indicative, in this regard, that Henrich does not investigate any of Heidegger's other texts of the time, and specifically misses or ignores the question of 'Kant's thesis about being'. Heidegger has given us a *topos* where the idea of a fundamental ontology may be tested: if it has or has not the power to illuminate. He radically criticizes Kant's addiction to the traditional doctrines of consciousness, logic, reason and imagination, and attempts to retrieve an *unsaid* possibility of the Kantian text – perhaps to rescue 'Kant' from the (Neo)-Kantians. In this context, Heidegger's goal is that of 'counter-ruination', a resistance in the midst of a prior situation of 'ruination'.

Ecstatic temporality and transcendental imagination

If the question of human pure 'reason' is that of finitude, then, for Heidegger, the problem of metaphysics is the question of the '"specific" finitude of human subjectivity . . .'[338] The 'specific' refers to an attempt to articulate a *topos* for an indigenous self-interpretation. This self is not some 'instance' of the 'finite rational creature in general' – instead, we seek phenomenological specificity of this existence of the self in his or her world. Pure imagination as pure *sensible* reason gives rise to such a finite knowing. In that pure sensibility is 'time', we are seeking to disclose a 'unity' of original temporality. Heidegger intimates a solution: 'Time as pure intuition is the forming intuition of which it intuits **in one**.'[339] An 'origin' of time is disclosed in pure imagination; time as pure intuition gives to itself that which is to be disclosed. Time has a 'free moving character . . .' It is not possible to freeze this present moment, 'now'. Pure imagination projects these horizons of discernability, of a 'likeness-taking' (present), reproduction (having-been) and prefiguration (future). Pure imagination 'forms a fabric out of itself',[340] one that is textured as time as such. 'It' gives time. Heidegger, in this light, re-names imagination:

> On the contrary, the transcendental power of imagination allows time as a sequence of nows to spring forth, and as the letting-spring-forth it is therefore original time.[341]

Pure imagination is the 'ground' for each synthesis; it points to original time as a unifying, though, finite 'fundament'. Heidegger admits that Kant resisted the pure imagination as a formative centre of the three syntheses as imagination. For Kant, imagination is tragically sublimated to reproduction, sensation and subordinated to a timeless 'unity' of 'reason'. It is the post-revision Kant who abides in the eighteenth century, one could argue. Even if Heidegger can point to 'susceptibilities' in Kant of his radical phenomenology of original existence, he is pointing across – into 'nothing'. His predicament uncovers the character of his solitude, as an 'artist of existence', disclosed in his confession:

> That the transcendental power of imagination is the root of sensibility and understanding was admittedly first shown in the more original interpretation.[342]

With this admission, we can begin to consider the temporality of the syntheses of imagination in more detail, and begin to approach a regard towards Heidegger's interpretive *praxis*. In this way, we will not be simply rejecting 'Heidegger's interpretation of Kant', nor will we be advocating 'caution'. After all, it is the transcendental imagining, as *ens imaginarium*, that provides the clearest link between Kant and the ecstatic projections upon the no-thing of Heidegger. Kant's 'thesis about being' indicates a projection of being, of existence, which discloses the meaning of his 'highest principle'. For Heidegger, saying it all differently, a horizon of being-in-the-world opens up the possibility of encounters with beings 'in' the world. It is a receptive spontaneity, and *vice versa*, which opens a place for the entry of beings.

The temporality of the syntheses of pure imagination

Pure synthesis as pure apprehension

The diversity of experience is dependent upon the differing or stancing of time. The separation of time into diversity (now this and now that) is the prerequisite for the de-severing which 'runs through' (*synopsis*) and 'collectively takes up the impressions'[343] (*synthesis*). This taking up is a forming-from and comprehension of phenomena, gathered, as we have seen, as a *look*, which offers itself 'at once' amid a manifold of experience. It is pure intuition, original time, pure apprehending *syndosis*, which 'forms' the look of the 'now', of the 'present', in a 'moment of vision' (**Augenblick**), as an original receptivity which 'produces' the 'actual present as such'[344] and its direction. Synthesis is time-forming; it forms the 'now' which is the domain of empirical intuition. The synthesis thus 'has a temporal character'.[345] Heidegger refers to Kant (A120) to the effect that imagination has an immediate synthetic operation upon perception, or, in other words, that imagination is intuitive apprehension. In this light, in that synthesis is time-forming, transcendental imagination must have an inner temporal character.

Pure synthesis as pure reproduction

Reproduction is a bringing forth again of that which manifests itself in a cycle of repetition. That which is brought forth into an awareness, however, need not be present, as the imagination can bring forth apparitions, looks, of that which is absent, of that which is 'not', as with Heidegger's 'hallucination' of a automobile flying across the classroom. The synthesis of average imagination, intentionality, maintains this power of unifying, if the 'look' is remembered. Memory is dependent upon a prior stancing of time 'earlier' and 'at that time'.[346] The no-longer now is retained by being synthetically joined to the current present unfolding, of a duration from then to [now]. Pure reproductive synthesis, which Heidegger calls a 'square circle', would be, in this way, the condition of possibility of an empirical reproductive synthesis. It 'brings the horizon of the earlier into view and holds it open as such in advance'.[347] In that it projects or forms the horizon of 'having been-ness', synthesis is time-forming, a comporting, a possible attending to 'having been' as such. Pure imagination is disclosed as a specific origin of time, time-forming with respect to 'having been', and, this here and now, of the now and the no-longer-now.

Pure synthesis as pure recognition

Recognition is 'consciousness' of that which remains the same between the 'nothings' of the no-longer-now and not-yet-now. 'Pure consciousness' serves as a ground or horizon for reproduction and apprehension. But, where is the criteria for this 'unity' amid reproduction, 'if what they want to present as unified and the same is, so to speak, placeless',[348] if that which was, becomes abandoned, displaced by the 'visualization' of a present, a duration that is *not*, and, as with any now, cannot be fixed there as a 'present' 'here' at hand? Yet, again, this duality of the now and no-longer-now, as in the case of the stems of knowledge, deceives us as we seek that which 'joins these syntheses . . .' For his part, Heidegger indicates an original unifying orientation 'as something which has presence in sameness . . .'[349] It is one 'consciousness' which unites the differing syntheses; it is the conceptual representing of a many, finding agreement in 'one' (A103). This 'one' is re-interpreted by Heidegger as the pure synthesis of 'identification', as a synthesis 'anticipating', 'reconnoitering' that which holds this 'nothing' before itself as the same, thus de-limiting possibility as the temporal ground of a 'circumscribed field of beings'.[350]

Pure recognition, as a pure synthesis, serves as the *a priori* basis of 'identification'. However, this event of identification, Heidegger indicates, is orchestrated within temporal horizons. Anticipation, reconnoitering, in the sense of a turning-toward a horizon of transcendence '. . . explores the horizon of being-able-to-hold-something-before-us [**Vorhaltbarkeit**] in general.'[351] 'Measure' is futural, it is an *a priori* horizon, awaiting the entry of beings as a prefiguration for that which is 'next'. Heidegger writes: 'As pure, its exploring is the original forming of this preliminary attaching [**Vorhaften**], i.e., future.'[352] Heidegger contends that as with the first two syntheses this third, that of recognition in the concept, is also time-forming. This 'pure preparation' is a mode of the transcendental power of imagination. But, what relation, if any, can apperception, allegedly timeless, have with time?

It appears 'absurd', Heidegger baits, that pure thought/apperception should have an 'essential' relation with time. But, in light of the reduction of the stems to the root, of an inner temporality of pure imagination, and the presupposition by thought of a synthesis of imagination, then, as Sherover mused, it would seem that Heidegger was only making obvious conclusions. He acknowledges the problems surrounding conceiving of 'the self as inherently temporal'.[353] Yet, he suggests that such a path of exploration is a 'possibility', not only with respect to the many remarks made by Kant and the many susceptibilities of the text of Kant, but also, and at the very least, in that temporality, as an original receptivity, pervades the character of the self as 'inner sense', of the 'subjective character of time'.[354] Time is at the heart of a self-disclosed as pure self-affection.

Time as pure self-affection and the temporal character of the self

Time is not merely a horizon for empirical sensation, but affects the self in its be-ing. Yet, 'where' is this affection, if not on the 'outside', if not 'present' in a being which 'announces itself?' How does time affect that concept, apperception, or the 'self'? Heidegger writes that original time, or derivatively, 'inner sense' in Kantian language, forms a succession of 'nows' from out of itself. The self 'seizes' hold of this 'succession', '*clutches* [this succession] as such *to itself*'. This temporal flow is its own receptivity, its 'formative taking-in-stride'.[355] In this way, pure intuition, Heidegger continues, projects the pure horizon through which the finite self is affected *by itself*. Time as pure self-affection unfolds the intentional nexus of lived existence. The opening of this being-in-the-world, being-there, incites a reflexive hermeneutics of this self/world. Time **is** this self. Kisiel translates Heidegger, in his Aristotle lectures, 'time is its own norm'[356] – temporality expresses/interprets itself. It is a self-activating, self -appropriation, a having of self, it is the 'essential structure of subjectivity'.[357] It is selfhood characterized by an inherent receptivity. Or, in Heidegger's language, it is a pure sensible (human) 'reason'.

The 'grounding' of transcendence, apperception, in original temporality, from the perspective of the proximate existence of *my* self, is an appropriation of that in which I am already involved. For Kant, it would be an acceptance of horizon of possible experience. Yet, Heidegger seeks to take the limitation of reason to its radical end: 'Pure taking-in-stride, however, means: becoming affected in the absence of a theoretically constructed experience, i.e., self-affecting.'[358] Original time thus 'forms' transcendence. In this light, that which most characterizes apperception, this letting-stand-against-of, the opening of an intentional relation, turns out to be original time, pure self-affection. The concept traces its origin to time, and discloses a kinship with imagination. Time as pure self-affection is the 'transcendental, primal structure of the finite self as such', a being-on-the-outside, a turning-toward and a return, back to one's self, one comes towards oneself amid an intimacy of self-interpretation and expression. Heidegger claims that this reflexivity is prior to inter-subjectivity (but not **Mitdasein**), as being is in each case 'my own'.

Heidegger, as we can see, has rejected a point of departure in the question of the relation of time and apperception. One is not alongside the other, but both are aspects of the same, recalling once again the 'many', differing ways of saying, of naming, of indicating the myriad fluidity of a 'nothing' that is the upon-which of our projections. Time, for Heidegger, *is* apperception, as each is a different name for the temporality of existence and thought. Original time or apperception (the 'I') forms the pure look of any 'present in general', '. . . the letting-stand-against-of . . . and its horizon.'[359] Heidegger writes that Kant was oriented to a non-original interpretation of time, and thus, with some irony, is 'correct' in his prohibition of a temporal analysis of pure reason. Heidegger asserts that in the case of Kant, interpretive violence is necessary in order to 'wring from what the words say, what it is they want to say . . .'[360] A *destruktion* in this way, has the power to push the 'theoretical' back into its ground of possibility, even if such a procedure may appear in the first instance as arbitrary and violent. Yet, that which is disclosed in the excavation, amid the interpretive act, is the radical temporalization of thought. The laying of the ground occurs upon a 'ground' of time. This excavation of the temporal be-ing of the self is a '. . . retrieval [Wiederholung] of its own, more original possibility.'[361]

Ecstatico-horizonal temporality and transcendental imagination

In the last section of the Kantbook, 'The Idea of Fundamental Ontology and the *Critique of Pure Reason*', in a marginal note by Heidegger, included in the Fifth English Edition, he asks: 'what does this mean?'[362] He is referring to the statement: 'Dasein projects Being from itself upon time, so to speak.'[363] We have seen in Chapter 4 that ecstatic temporality occurs via projections, and that such a projection is a self-projection of Dasein, the *topos* of any temporalizing of original, ecstatic temporality. Specifically, we have seen that a schema is projected with each ecstasis as a horizon of possibility (as with *praesens*, the horizonal schema for the ecstasis of the present) and that it includes in its own projection an 'upon-which' (**woraufhin**) of projection for an understanding of being.

That which breaks in, intimates that which is beyond the familiarity of the being of the extant, is an event of self-finding, an irruption of unconcealment, free of homogenized absorption. Since 'nothing' is 'beyond' this being of the extant, real predication, and since it is not, as that which breaks in, nothing at all, it has a mode of being that cannot be understood only amidst the horizon of *praesens*, of the present, but must also unfold the horizons associated with the future and the past. It is in this situation that the note indicating the projection of 'being upon time' is raised. I reiterate: 'what does this mean?' Heidegger speaks in *Basic Problems* of the question of the meaning of being as such. Yet, such a determination can only be made, according to Heidegger, once we have described the ecstatic *topos* of orientation for the questioner, Dasein, beginning with being as everydayness, with the familiar. Heidegger states that the meaning of the being of the extant, as an understanding of being *per se*, is described as a projection of being upon time, and specifically in this case, upon a horizonal schema of *praesens*. The meaning of being of the extant, is, in this instance, presence: being is presence, but only in the sense of one schematic horizon of original ecstatic temporality. This meaning of being is distinct, for Heidegger, from that of other Praesens which is the meaning of being *as such*, and of which he begins a discussion, but does not finish, in *Basic Problems*. However, it could be argued that in light of the temporal problematic, the meaning of Praesens would acquire the character of ecstatic temporality, or that *Praesens*, as the meaning of being, would abide and give rise to aspects of the present that are *not themselves of the present*, but bleed into the secure island of the 'now'. In other words, we must attempt to set Praesens free and continue upon the path of Heidegger's discussions by means of an elaboration of the complexities of ecstatic-horizonal temporality with respect to the meaning of being. As we get a taste in Heidegger's later lecture, 'Time and Being',[364] 'Praesens' would thus be the playspace of ecstatic temporality itself amid its three ecstasies and their respective schematic horizons. The projection of being, or, an understanding of being upon the schematic horizon of an ecstasis of original temporality, unfolds with an inextricable 'unity' of the ecstasies of temporality which allows us to begin the separation between *praesens* as the horizonal schema of the present and Praesens as the horizon for Being as such. In this event, Praesens could begin to be interpreted in a differing way which would not fall prey to the fallacy of 'presence' and its metaphysically loaded lexicon. The path of Heidegger leads to an unfolding of the complexity of the three ecstasies with respect to the articulation of their respective horizonal schemas.

Heidegger did not move beyond the horizonal schema of presence in *Basic Problems*, though he did obviously discuss the future and having-been-ness at great length in other contexts. In this light, there is need of a provisional indication of the other horizonal schemas, rooted in the future and having-been, and in their mutual interaction amid the event of being as the playspace of ecstatic temporality. We will thus indicate the other horizons of ecstatic temporality, from Latin in the manner of Heidegger in the case of **praesens**, with the words **futurens** and **passens**, from *futurus* (husteron) and *passus* (proteron). With these schema brought into play, there is the possibility of moving beyond the horizon of the present only so as to understand Praesens as the 'unity' of the three ecstasies. With the elaboration of the horizonal schema of futurens and passens, this indication of *nothing*, of *ens imaginarium*, becomes elaborated as a triune ecstatico-horizonal schematism. The ecstatic schematism, or ecstematic, indicates a threefold of projections and the *upon-which* of these projections. It finds its unity in the 'makeshift' resolution, in the temporal event of intimate self-interpretation in which it would be possible to disclose one's own Being.

Amid this multi-directional temporalization, the horizonal limit of the three schema of ecstatic temporality disclose the *topos* of projective understanding. Understanding is a projection, a throwing forth, a casting, a spontaneity. Yet, at the same time, it is a finite, thrown projection, dependent upon this receptivity of the givenness of the being. In this light, this ecstatic *topos* indicates an interpretation of a complex temporality in comparison to that allowed amid the horizonal schema of *praesens* only (that is from amid the horizon of the 'present'). As such, in the self-projection (**Entwurf**) of ecstatic temporality, if we take into account each of the three ecstasies, each a double projecting, there will be no less than six projections which constitute this 'self/world' amid any given instant. Such an interpretation of time will allow our understanding of being to dispel its captivation to a picture of presence, and to become oriented amidst the makeshift play of the ecstases of temporality, where anxiety and boredom are modes of finite, sensible understanding.

Kisiel suggests that such an ecstatic *topos* is a 'middle voiced phenomenon', a place of 'being troubled and troubling oneself', in a 'unity' of past, present and future.[365] The unfamiliar expressions of projection and 'upon-which' are meant to indicate the intentional place of spontaneity and receptivity. Yet, such an awkward expression speaks again of pure imagination, the original root for these stems of spontaneity and receptivity. As we have learned, imagination not only mediates; it originates the stems, it is existence as self -affection, as temporality. Of course, there are significant differences between his unfinished account of the ecstasies in the 'Sein und Zeit' project and his essay 'Time and Being'. And, these differences can be instructive in the specification of the operation of ecstatic temporality in the early Heidegger. In the essay, for instance, ecstatic or original temporality allows for a *topos* for the free play of the ecstasies. Yet, this seemingly directionless play of ecstasies, as the play of Being, must be contrasted with the priority of the ecstasy of futurity in *Being and Time*. In the latter text, time as pure self-affection refers to disposition (**Befindlichkeit**) and self-disclosure amidst 'world'. Dasein, existence, and its inevitable trajectory towards the possibility of impossibility, remains the focal point of the question of Being, just as the future remains the dominant ecstasy of time. With the turn away from the finite self in the *metontology* of the truth of Being, there remains only the play of Being.

Part 3

The *Topos* of Ecstatic Temporality

We wish to revive neither Aristotle nor that ontology of the Middle Ages, neither Kant nor Hegel, but only ourselves, that is to say, we emancipate ourselves from the phraseologies and conveniences of a present which reels from one fickle fashion to the next.[366]

If we wish to revive ourselves, emancipate ourselves, we must complete the circle, as it were, and set out the specific conditions for, and the enunciations of, an indigenous conceptuality. Yet, before we consider Heidegger's explicit unfolding of the existentiales in *Being and Time*, we will turn to his lecture course on Leibniz, 'Metaphysische Anfangsgrunde der Logick im Ausgang von Leibniz' (1928), which was to be *his last* at Marburg (and just after the lecture courses *Basic Problems* and *Phenomenological Interpretation of Kant's Critique of Pure Reason*). It is interesting that there is no indication of Leibniz in the projected outline of the 'Sein und Zeit' project. We will find, however, that Leibniz will take on important significance in light of his notion of a monadic 'drive' (**Drang**), and for his attempt to articulate a *mathesis*, a context of orientation, in which the 'unity' of existence could be expressed.[367]

For Heidegger, the essential insights of Leibniz, as it was with Kant, but probably more so in the former case, remained unsaid (and I would argue that this reticence emerged probably less as a 'shrinking back from the abyss' as from a desire to not follow in the footsteps of Giordano Bruno). Heidegger states that Leibniz's philosophical expression, in its ostensible prioritization of the medium of logic and of the proposition, failed to disclose the primary temporal significance of the *entelechy* as the drive of the finite self, and thus, as the be-ing of transcendence. Heidegger has his sights on Leibniz's principle of sufficient reason, and, upon the consequent severance between the truths of fact and truths of reason. Yet, without merely setting these principles aside, he excavates the labyrinth of eternal truths in Leibniz and discloses the primordial dimension of a transcendental and temporal self amid existence. Tracing the 'common root' in the context of Leibniz allows us the possibility to consider Temporality as the meaning of the principle of ground, as that ecstatic playspace which is the *topos* of transcendence. This entering into 'Leibniz' with respect to the 'relation' between being and *logos*, and thus, of the temporality of thinking, will lay the ground for the third, *constructive*, aspect of the 'Sein und Zeit' project. In this light, we will be exploring not only the phenomenological and existential conditions for self-interpretation/expression, but, will also explore the being of the interpretive/expressive 'act' itself, that of the question, as this fruit that is grown from a branch, stem, trunk and root, but, is itself, at the same time, free and forced to fall away from the tree.

An honest phenomenology, for Heidegger, would be one of Care (**Sorge**) which seeks, with minimal intervention, construction, *this* self-expression of the phenomenon. Such a movement, shift, prepares the 'ground' for the 'event' of self-interpretation and expression in the 'moment of vision' (**Augenblick**). Questioning, as a nexus of indications and transfers, projects horizons of significance, in which the phenomenon can express itself, uncovered, be brought out into the clear light. Yet, as we have also seen, the expressive 'act' gives rise to the possibility of a *logos* that is free-floating. Heidegger works hard to articulate existentiales which remain rooted and attuned to the phenomenon. Yet, as each moment expressed is a new translation (Heidegger's lecture course on Parmenides and Heraclitus, Winter 1942–43) amid the *topos* of ecstatic projection, even such 'dedicated submission' must itself be open to its own makeshift status. At the same time, Heidegger also expresses the possibility of *silence* amid the ecstatic transcend-ing of the self, as it holds itself in the no-thing, to be broken and brought to the unconcealed manifestation of one's own finite possibility.

Chapter 10

The Metaphysical Foundations of Logic

But instead of embracing this unleashing of oneself, a being stops in himself the torrent which gives him over to life, and devotes himself, in the hope of avoiding ruin, in the fear of overpowering glories, to the possession of things. And things possess him when he believes to be possessing them.[368]

Between the publications of *Being and Time* and *Kant and the Problem of Metaphysics,* Heidegger gave a lecture course entitled 'Metaphysische Anfangsgrunde der Logick im Ausgang von Leibniz' (1928), published in English as *Metaphysical Foundations of Logic* (1984). In his last course given at the University of Marburg, Heidegger deconstructs logic within the horizons of the question of a *logike episteme* (λογική έπιστήμη), translated traditionally as 'metaphysics of truth'. With the interpretation of truth as *a-lethea,* truth as uncealment is neither a bridge towards an object from a subject, nor as a crossing over towards a divine. It is instead a surpassing, an excession, or, movement beyond beings in the projection of an *a priori* horizon, which we have discussed above. The horizon transcending is an anticipatory 'sketch' (**Entwurf**) in the context of which beings will be disclosed. In the first half of this lecture course, the subject matter for this chapter, Heidegger begins a dismantling of 'traditional logic' so as to allow for the expression of free, finite thinking. He remarks that logic does not 'speak to the students', since it is uprooted from its 'metaphysical' origins, severed from its historical enactment as expression, as an utterance of *logos.* Heidegger charges that 'logic' is an 'unworlded' syntax, which not only suppresses truth and its indigenous expression, as a *topos* of original temporality, but also prohibits the question of the origin of logic, which after all is allegedly free. *Logic,* for Heidegger, is merely a semblance of thinking, with an endogenous 'logic' alien to the matters. Yet, he does not seek to annihilate logic, but instead, to retrieve an original sense of a temporal *logos* that uncovers the original *topos* of singular expression. A *destruktion* excavates the temporal *logos,* and releases it from its 'petrifaction', its dead 'stuff'.

In the following pages, we will examine Heidegger's dismantling of traditional logic in tandem with his desire to express an honest philosophical *logos,* an expression arising amid the event character of temporality, as indicative characters of be-ing. In other words, we will explore Heidegger's retrieval of a deeper, original sense of *logos*, along the path of the question of 'ground', of the limits, or rules, for thought. Such limitation of thought by rules will itself be interpreted in light of the limitations inherent in the notion of a pure, *sensible* reason, of a finite understanding of being, which we have explored in Part 2. In this context, we will ascertain that a deconstructed 'logic' points to a 'ground' of finite, *a priori* knowledge, transcendence, as an existential freedom of the 'thrown dissemination' of being-in-the-world. In this way, we will cast into relief the expressive element of *construction* which, as with *Basic Problems*, is the third aspect of the phenomenology of original temporality, as an articulation of self-understanding.

A question of philosophical logic

Heidegger, in a question he would later echo in his 1929 Inaugural Address at the University of Freiburg, 'What is Metaphysics?', asks if it is not absurd to throw into question the bi-millenial tradition of logic, to shift its foundations, simply so that it may speak to us, that it may become an indigenous *logos*, an expression of temporal singularity. Yet, through his ironic and rhetorical use of the word absurdity, he will attempt to overcome the stigmas which become attached to strategies of questioning which seek to disclose the temporal and metaphysical 'ground' of logic. Heidegger begins his investigation with a genealogy, tracing the emergence of logic within the horizon of the history of philosophy, in its 'gigantic beginning' with the 'Greeks', indicated in the phrase, *logike episteme* (λογική έπιστήμη). He will seek to retrieve an original sense of 'logic' from an excavation of the historical and temporal facticity of 'logic', one that will uncover a deeper 'place' of questioning in the practice of intimate self-interpretation.

The point of departure for Heidegger's *destruktion* of 'logic' consists in the characterization of the meaning of philosophy from out of the horizons of its own historicity:

> Philosophy can be characterized only from and in historical recollection. But this recollection is only what it is, is only living, in the moment of self-understanding.[369]

Heidegger advocates neither an antiquarian preservation nor an elimination of the history of philosophy. Neither does he undertake to install himself into the philosophical questioning of another, in the manner of the scholar, and from this standpoint, to commence to 'tailor it and supplement it'. What is 'necessary' is a 'moment of self understanding' in which one has entered into the question through recollection, and has appropriated this original questioning, not as a 'pale' and self-conscious repetition, but as a retrieval of an *eigentlich* self-interpretation. Such a 'recollection from historicity' is one's own free projection of questioning and as such, a projection which is an ecstasis of one's own original temporality. It is, in this way, that 'recollection' exhibits some peculiar aspects with respect to the event of temporality. From the word itself, in its common usage, we may imagine that Heidegger is a conservative philosopher of the past, of tradition. Yet, such an interpretation immediately comes into conflict with Heidegger's radical project to *destroy the history of ontology*. This conflict forces us to understand the meaning of recollection and our own relationship to the past in a different light. Such a differing light may cast into relief Heidegger's enigmatic claim that, in the context of original, ecstatic temporality, the *having-been-ness* of the past is a temporalization which arises out of the future. This claim could make no sense at all in the context of 'common time'. Yet, with respect to the projection of the original horizon for the entry of beings, as an *a priori* horizon which opens in the anticipation of one's own death, it is not difficult to understand how the future could precede the past. In this sense, historical recollection, or the appropriation of that which has been in a projection of one's ownmost possibility, will be the way by which we disclose the original sense of *logos*. Of course, we do not begin from scratch as we already, in our comportment with history, possess a pre-understanding of *logos*. Admittedly, to some extent, we move in a circle, but as Heidegger has suggested, that which is important is entering into the circle in the 'right way'. In this context, the 'right way' would be to understand that recollection, as a way of comporting to the past, is only what it is as a moment of self-understanding, as the projection of futurity. In other words, historical recollection is the projection of one's ownmost possibilities amid the ecstasy of the future.

Taking a few steps back, Heidegger begins his historical recollection with an exploration of the etymology of *logike episteme* (λογική ἐπιστήμη), translated by the tradition as the *metaphysics of truth*, a science of the *logos*. Heidegger sets out his interpretation of the 'science of logic' in light of the original meaning of *logos*, in the sense of speech, statement and predication: to say something about something, *determinatio*, or, he states, 'thinking'. That which is said, the statement, has a 'relationship' to that about which it is uttered. The statement is made about the being itself. The relationship between the statement and the being is the 'in-between', as it were, the *topos* of a simultaneous determination and disclosure of a being. The statement, conceived as an indication, exposes its own truth or falsity, in its 'measuring up', by its adequation, with this being. Heidegger, sounding like Quine, states that 'logic' cannot analyse the truth or falsity of all statements, however, and must therefore orient its own examination to the properties of 'statement as such'. Yet, as a criticism of the *logicians*, he states that a search for an originary *logos* must not overlook the ontological difference or the diversity of regions of phenomena, together with their indigenous conceptualities. This is an echo of Heidegger's demand that phenomenology remain oriented to the 'matters themselves'.

As an excellent example of his *destruktive* methodology, Heidegger seeks to corroborate these criticisms with an interpretation of 'general logic' that reveals the 'ontological difference'. As general, he states, it must not be oriented to beings here and there. Indeed, to be honest, general logic is concerned with *nothing* at all. General logic is neither a 'material logic', nor a circumspective phenomenology of extant beings. Logic, as it is not concerned with beings, with the ontic, is, Heidegger declares with a clear sense of irony, a phenomenology of nothing (**das Nichts**). Logic is thus 'thinking about nothing'. This indication of no-thingness, a term which would later provoke horror for Carnap, is an intimation of the *prius* of transcendence that allows for possible encounters with beings here and there. 'Logic', from this perspective, is an interpretation of transcendence, and thus, is a 'hermeneutics of existence'.

In its transcendent interpretation, general logic is concerned with 'lawfulness', or formal rules of thought. Heidegger contends however that the sense or meaning of 'formal rules of thought', specifically in Kant, remains unclear with respect to the difference between 'formal truth' and 'correctness'. In this situation of unclarity, 'logic' as *logos* falls away from its expression of an *a priori* horizon of discernability and becomes entangled in an entitive metaphysics that merely polices the identities of things. Heidegger declares that there is 'need' for a 'new logic'. What is idolized as 'logic' is 'not a logic at all'– it shows 'nothing in common anymore with philosophy'.[370] This new logic, as the openness to a 'ground' that is no ground at all, is oriented to *a-lethic* truth, a philosophical 'logic' that facilitates 'radical possibilities of questioning'.[371] Amid this 'dadaist' rift betwixt judgement and truth, Heidegger's *destruktion* begins to dismantle the familiar idols which have captivated us, and thereby makes the 'obvious incomprehensible and the unquestioned something questionable'.[372] It is not a matter of denying the 'validity' of 'laws of thought', but to expose their derivativeness in the face of 'original matters'.

Instead of proceeding from an extant definition of philosophy, and then, to a derivation of logic, with the consequent exposure of a dependence by origination, Heidegger seeks to enter into this 'logic', to 'understand' by inhabiting its basic situation of questions and problems. If 'logic' does indeed trace its origins to philosophy, these will be, he contends, philosophical problems. It is only in the *destruktion* of logic to its original state, that 'logic' can be a propaedeutic, or intro-duction, to philosophy. At the same time, such an entering into 'logic' must be, from the perspective of Heidegger's project, a 'recollection from historicity'. In a provisional sense, philosophy has been indicated as a self-interpretation of finite or historical existence. In this way, if we are to enter into the questions of logic, we must do so through a specification of these questions through a *destruktive* reading of a historical philosophy. Heidegger, in this light, specifies the historicity of logic through an examination of Aristotle's definition of philosophy, as that practice which is both a pre-philosophical question of existence and a theo-logical apprehension of the overwhelming. This recollection from history is vital in light of the influence of the Scholiasts upon Leibniz, and also with his own appropriation of Aristotle. It would be well to remember throughout the following, the implications of Heidegger's analysis in light of his ongoing battle with the Vienna Circle.

Aristotle's definition of philosophy

Heidegger casts his 'return to the Ancients', particularly to Aristotle, as a needed retrieval of an 'elemental originality' of basic philosophical problems. Echoing the contention of early phenomenology of the priority of lived understanding against mere 'known knowledge', his retrieval of the simplicity of Aristotle's expression of the basic problems is characterized by him as a 'present necessity' in the face of the 'disoriented psychologizing chattiness of contemporary philosophy'.[373] Aristotle has been chosen as he is regarded by Heidegger as the peak of ancient philosophy, although neither its fulfilment, nor its own depth, or unmined possibilities. The philosophy of the historical Aristotle is, in this way, only a 'makeshift', a free projection, a sketch (**Entwurf**) of his being-in-the-world.

Aristotle's definition or sense of philosophy is expressed as the question of being, articulated in the *Metaphysics* as: 'There is a definite science which enquires into beings as being and into that which belongs to it as such'.[374] The question of being is designated as *philosophia prote* (φιλοσοφία πρώτη), as 'first philosophy', of beings as beings, in their being, as *to on e on* (το ον η ον). Yet, the meaning of being is not clear, and this predicament of unclarity indicates an array of basic problems at the heart of *first philosophy*. With respect to the meaning of being, we can proceed at first only *via negativa*, that is, being is not beings. In this light, Heidegger contends that we cannot ask of a 'what', but of *that*, of a no-thing, which, as we have indicated, is the transcendental horizon for existence. This latter horizon is intimated in Aristotle's definition of philosophy which includes a *theologike philosophia* (θεολογική φιλοσοφία), where *theion* (θειον) means 'simply beings – the Heavens'.[375] The domain of this inquiry is that of a 'first', kosmos (κοσμος), which is 'beyond being', and for Heidegger, encompasses that 'under and upon which we are thrown, that which dazzles, surprises, shatters our world, as the overwhelming'.[376] As we have seen, the many meanings of being are given, as temporality and its expression, beyond being. In this sense, the overwhelming is that which incites the disclosure of the finite horizons of one's own existence. And, in light of the radical temporality of such 'thinking', Heidegger contends that *philosophia* (φιλοσοφία) must be an *episteme zetoumene* (ἐπιστήμη ζητουμένη): a finite knowing is sought, never to become the 'fixed possession' to be 'passed on'. Each 'first philosophy' must begin anew, as a trusting concern, love (φιλεῖν), desire towards an original understanding (σοφία) of 'matters themselves'. 'Desire', or Care in *Being and Time*, is, as we will see in Chapter 12, the meaning of the be-ing of Dasein, but this means that Care is projected upon amid an original, ecstatic temporality, beyond being.

Finite knowing, Heidegger continues his analysis, is concerned with *peritta* (περιττά), or, that which turns all upside down, and with *thaumasta* (θαυμαστά), the wondrous that provokes an original questioning, *chalepa* (χαλεπά), the difficult, and *daimonia* (δαιμόνια), that which holds humans in thrall. He writes significantly that such concerns are *achresta* (ἄχρηστα), useless for 'day to day necessities'.[377] In this way, 'first philosophy', as the 'useless' wisdom that each time must be sought anew, cannot be merely circulated in familiar doctrines purchased at the local pharmacy. On the contrary, such knowing is gathered together amidst this desire, in this practice of *seeking-after*. It is an indigenous understanding, an entering into the question of the meaning of being in a free projection. That which is *sough-after* is knowledge of the 'first', as with the early Greek philosophers, prior in the sense of an *a priori* of beings. Such an *a priori* knowledge of being is a *logos* (λόγος) of the *on e on* (ον η ον); it is, to this extent, an 'ontology', which is concerned with the being which is that of the questioner *in singularis*. In this way, in that philosophy concerns that which is 'beyond being', it is also a 'theology of the overwhelming', an intimation of *this*, of one's ownmost possibility.

The question of being and the question of man

Heidegger comments that the question about being has been oriented, since Parmenides, to the questioning-after-being, and thus, for the one who questions, this questioning is the thinking of being. Such a shift to the field of thinking, which he associates with 'soul' and 'subjectivity', and thus, human Dasein, is possible since an understanding-of-being is the basic character of existence. This understanding-of-being most basically pertains to the being of Dasein, whose character it is to question after the being of the being that it is, to 'care' for its being in the world, its existence, in light of her or his own temporal predicament. Understanding, in this light, orchestrates the primal transcendence that is the prerequisite event for any comportment to beings. The event is an opening towards, or, to use Heidegger's language, a being thrown upon beings. In this way, the question of being, **the** question at the heart of human Dasein, is, at the same time, this 'question of man'. In other words, each question is an expression of the 'basic question of philosophy', between which there has been, Heidegger alludes, a hidden relation throughout the history of philosophy. To ask the question of man in light of, or projected against the horizons of, the question of being, is a provisional disclosure, of the unsurpassable limits of possibility for our existence. Such an 'analytic of dasein', as an exploration of limits, must not only be prior to the 'sciences of man', anthropology, sociology, psychology and medicine, for instance, but also, must be 'entered into' as an intimate hermeneutic of one's own existent self. Heidegger states:

> Human Dasein gains depth only if it succeeds for itself, in its own existence, in first throwing itself beyond itself – to its limits. Only from the height of this high projection does it glimpse its true depths.[378]

It is in this way that the radical phenomenological preference for the matters themselves is a 'loyalty' to the self against 'theory' and the non-indigenous, arbitrary and violent hegemony of a free-floating logic. That such a semblance of thought can arise amidst the detour into philosophical expression, intimates, for Heidegger, that a shift to thinking is 'dangerous' to the one who seeks after truth in the sense of *a-lethea*. 'Logic' has acquired the meaning of logistics, and is concerned with the 'truth of correctness'. The desire for truth in the former sense, however, is, for Heidegger, the intentionality and destination of our freedom, conceived as a primal transcendence, an understanding of being, disclosed in the moment of vision, in the *Augenblick* of self-understanding. The 'test' of the limits of freedom is in our capacity to hold ourselves amid this moment, to hold ourselves in the nothing (**das Nichts**). Such a moment amid finite knowing is intrinsically fragile, in the play of *a-lethea*, as truth and untruth, unconcealment and concealment, abide together as modes of the temporalizations of being (there) in the world. The question of thinking and of its laws, for Heidegger, becomes the question of the being of Dasein itself, and of its comportments that are the 'conditions of possibility' for these laws, and for this thinking. The existential conditions or 'basic principles' are the grounds not only for the statement, of logic as such, but are, at the same time, and more fundamentally, the constitutive modalities of human existence and its understanding of being. Heidegger, in other words, has uncovered the initial problem of logic, that of self-regulation, to be the question of the primal transcendence of finite, but irrepressible, human freedom. And, this is at once the question of truth, and of being in the truth, and as I have suggested, it is also the question of falsity, of a concealing which covers over 'matters' and remains entangled in entities. The intimacy of thought and comportment, and of truth and untruth, are, for Heidegger, constitutive aspects of the be-ing of Dasein.

With this provisional understanding of the meaning of philosophy, Heidegger turns to Leibniz, declaring that his work is a *topos* for a breakthrough into the basic questions of logic, expression and temporality. He contends that Leibniz expresses an original appropriation of the temporal problematic and the question of the 'metaphysical foundations of logic'. Leibniz is a place, moreover, where ancient, medieval and modernist interpretations of 'initial logic' (**initia logicae**) converge. This original sense of *logos* indicates an original 'unity' of

voice, and is associated in Aristotle with lived expression, with breath, air. Such
an indigenous *logos* indicates a complex situation of aspects; not primarily an
external linking, an aggregate of severed notions, but is, in the originary sense
of the primal category, an indication of the 'unity' of a being. Statements either
affirm, *kataphasis* (κατάφασις) or deny, *apophasis* (απόφασις), but only as determinations that are guided by existence. Within the context of an original unity
of being, each element of statement, or *phasis* (φάσις), exhibits a meaningfulness that is projected amid the horizons of beings themselves, just as each
monad is the mirror of the entire cosmos.

Dismantling Leibniz's doctrine of judgment

Heidegger begins his investigation of Leibniz with a dismissal of the controversy
over the relative priority of logic or ontology, whether being arises 'from' thinking or if logic is grounded metaphysically in ontology. Again, he will reject
a point of departure in the question of a relationship between severed 'stems',
as these must, as we have seen, be bracketed or deconstructed, as to allow a
phenomenological interpretation of Leibniz's theory of judgement. Such an interpretation will excavate the pre-theoretical and pre-practical aspects of the
phenomenon of judgement in a *destruktion* of the theoretical formalism of its
expression. Heidegger will put it to work, testing its limits, so as to 'enter into'
an understanding of judgement via a retrieval of an understanding *in situ*. In
this way, from this perspective, judgement *as such* not only exhibits the character of statement, with a concern for 'truth', but also, that of comportment, as
a way of being (*ethos*). Heidegger points out that the traditional formulations of
the 'theory of judgment', those of Aristotle, Leibniz and Kant, hold that judgement is a *symploke* (συμπλοκή), connection (nexus, **connectio**), joining or a
synthesis (**compositio**) between two distinct concepts or representations. Yet,
he contends, these definitions of judgement harbour a 'disharmony', indicative
of, as Kant himself expresses, the absence of clarity with respect to 'in what the
asserted relation consists'.[379] As we can already see, the question of the *what*, of
essence, being, indicates the question of the ontological conditions of judgement, beyond a mere analysis of concepts and of their relations, to a radicalized
phenomenological ontology where there is uncovered a 'way of being' of *logos*
with respect to its own being-in-the-world.

   Heidegger, again returning to the 'Greeks', characterizes the relation of subject and predicate in a statement as that of a predication about that which
underlies, of the *hypokeimenon* (ὑποκείμενον), or **subjectum**. The statement is
said of that which underlies; it indicates the subject; the predicate term itself
is a 'sign' of the subjectum. A predicate is said to be 'contained in', included
in, the 'subject', although Heidegger warns, this saying is unclear with respect
to the *being* of an inclusion. This question of being arises in the context of an

excavation of the notion of 'containment', so as to retrieve from beneath a mere theory of inclusion, an original sense of *inesse*, a modality of *esse*, being. In-esse, or in-being, he tells us, is a 'containment' of a predicate 'in' the concept of a subject. The latter, of which this predicate states something, is the 'individual substance', or the monad. Substance (ουσια) is an 'independent presence' which can never be 'named nor said of' another ὑποκείμενον. The predicate is, on the contrary, as that which is common to many individuals, 'essentially not independent'.[380] The predicate is thus contained, included in a subject either explicitly, as an identity, or virtually, as a hidden, *techte* (τεχτε) identity. In this way, *a priori* knowledge will express the *totality* of predicative determinations, or identities, which are contained in the subject. Or, the meaning of truth is inclusion in the subject,[381] and the totality of its predications of inclusion is 'Reality'.

Heidegger states that Leibniz gives most ground to the logical relation of a subject and predicate, which indicates a priority of a logical over an ontic conception of the subject, or monadic substance. In this way, the existence of an existing being has been 'assimilated to the concept of the subject'.[382] With the question of being, and thus of the ontological difference, Heidegger states that the diversity of aspects of judgement, of the logical and the ontic, must be clearly distinguished throughout the dismantling of the 'theory of inclusion', even if each will be shown to be, in a peculiar way, coincident. The dismantling of the theoretical controversy between logicism and ontologism gives space for the question of the relation between inclusion (**includi**) and being-in (**inesse**), which in turn will clarify the relation of a philosophical logic and ontology. As suggested, the distinction is ultimately nil. Being true is the same as being-in. The simplest statement of inclusion is one of identity. From these criteria, true statements are adequate, real connections (**nexi**) of identity. Or, being true is either being self-identical or a being the same as (**idem esse**) that which is identical to itself. To this extent, the questions of truth and of being are 'intertwined.'

Heidegger points out two senses of truth in Leibniz, original and derivative. That which is original is that which is immediately evident, as in the Kantian analytic judgement, while that which is derivative, or, synthetic, must be analysed in the attempt to reduce it to original truth as identity. That truth which reduces to identity is moreover necessary, while that which cannot be so reduced is contingent. These differing senses of truth, original and derivative, Heidegger suggests, correspond to different types of beings, necessary and contingent beings, or, to that which is uncreated and the created. Each of these 'regions' of beings correspond to a temporal differentiation of eternity and time, and each is associated with a specific mode of truth: for the necessary being, a truth of reason, and for contingent created beings, truths of fact. The exemplar of an Absolute being and its knowing, as it alone is capable of reducing time to eternity, intimates an 'ideal' knowing. This 'ideal' is that of the intuitive understanding of 'God' in a *nunc stans*, an absolute 'self-consciousness'. Contingent 'facts' contain a trace of the absolute intellect, although one embroiled in the volition or will of 'God'. It is necessary truth which is ruled by the divine intellect alone. In the divine panorama, each facticity is anticipated in the silence of an infinite knowing 'beyond' the propositional modula of *symploke*, of synthesis and diairesis, of a 'serial discursiveness'.[383] The realm of facticity, Heidegger continues, is governed for Leibniz by the principle of sufficient reason, which itself is grounded upon the principles of contradiction and identity. Primary, original truth, as we have seen, is an express and simple identity, which is without reason. Secondary, but, necessary, truths are those which can be reduced to an identity, implying non-contradiction. In this way, these truths are subject, unlike the primary truths, to the principle of ground, or, to the principle of reason which for Leibniz 'holds first rank, albeit unclearly, among the principles'.[384] Heidegger claims that the *principium rationis* was not clearly explicated in Leibniz, but merely shelters under this 'sign' a collection of various principles and associations, each exhibiting an affinity to a myth of absolute knowledge, a *scientia Dei*. The knowing of the divine intellect is an immediate intuition of all things, past, present and future, in a pan-opsis of eternity. In this context, knowing is intuition, which, in this case, is an ever-lasting intuition of an infinite knowing. This latter *totum simul* becomes the 'paradigm' for knowing as such, and eternity, 'inferred' from simplicity and immutability, becomes the 'measure of time' (**mensura temporis**). Direct intuition becomes the paradigm of truth. It is with respect to this 'measure' that we have delineated other possibilities of truth. Less than direct intuition, inadequate knowing, is either obscure, confused, blind – each of which is a stage in that analysability of a notion 'into marks and moments of marks (**requisita**) managed to the end'.[385]

Leibniz, the mathematician, insists upon the intuitive, absolute knowing of a completely analysed notion, even if only as 'cognitive ideal'. However, it is this 'Leibniz' which Heidegger seeks to dismantle back towards an originary state of questioning and self-interpretation. He seeks to displace, amid the temporal situation of existence, Leibniz's *mathesis universalis* (universal order), and of its 'ideal' of infinity, as the *Gestell* for human, that is, finite knowing. Leibniz, with his 'truths of reason', to which even God is subject, nullifies, for Heidegger, the meaning of the difference between the finite and the infinite. It is his obsession with logic over being which allows him to be blind to the finite horizons of existence. Indeed, Heidegger uncovers a tacit link between the notion of truth as identity and that of 'being adequately perceived by intuition', a link which exposes how the philosophy of Leibniz became incarcerated by logic. Truth as intuition envisages a 'coherent connection' of a thing compatible, attuned with itself, and which is thereby, according to its essence, possible as a 'harmony of multiplicity'. It is this harmony of multiplicity which, Heidegger charges, is usurped by the nexus of judgement, the proposition, and of its concomitant interpretation of the law of identity as a free-floating, unworlded 'rule of thought'. For Heidegger, the harmony of multiplicity must remain attuned to the harmony of comportments since identity need not be conceived in the sense of 'logic', as a homogenous, a-temporal uniformity, but can instead be understood, as he indicates in his 1927 lectures, *Phenomenological Interpretation of Kant's Critique of Pure Reason*, as an event of 'identification', as a coming to a stand of a coherent, compatible 'thing'. Heidegger states that Kant himself became incarcerated by logic in his link of truth and identity in the 'primordial unity of the synthesis of transcendental apperception', and thus, in the 'logical subject' which is the 'I' that nominally makes statements as such. But, as we have begun to understand, such a logical subject must always already be related to an ontological-metaphysical subject, as is 'logic' to being. It is the task of the *destruktion* to disclose this relation.

The root of identity and being: *Monadology*

The subject of a statement, is, for Heidegger, at the same time, the individual substance, a being. Amid this 'monadological' sense of being, he asks after the 'bridge' between being as identity and being as monad. He states that the latter term comes from Greek, *monas* (μονάς), through Giordano Bruno, and refers to the simple, unified, the one, and to the individual and to the solitary. A monad is that which unifies, individuates in advance as an individual substance, and it is the *a priori* in Leibniz's 'Metaphysica Generalis'. Contrary to Descartes' identifying of *res corporea*, bodily thing, with *res extensa*, extended thing, an individual substance or monad is a positive, active and unifying 'primitive force' (**vis primitiva**). Primitive force inhabits a metaphysical, distinct from a mathematical point. Mathematics, for Heidegger, is merely a negative delineation of the 'limits' of a 'thing', and as an infinite fragmentation *without* limits, it cannot give 'unity' to the 'continuum' of being. The monad requires a specific modality of expression and understanding, distinct from that of mathematics. In this way, Heidegger contends, Leibniz has given back to philosophy its own concept of force, *vis primitiva*, a concept that not only anticipates Dilthey's 'life philosophy', but also, phenomenology.

The background for this restoration of a philosophical concept of force is, contrary to many such as Russell, who would see Leibniz as an analytic philosopher, the Scholastic distinctions with respect to the concept of power. Leibniz risks associating with Scholastic 'substantial forms', so as to express a concept of *active force* as the *modus operendi* of the individual substance, or monad. Heidegger traces this notion of active force to Scholastic, specifically Aquinian, concepts of passive and active power in its relation to the notion of 'act', as explicated in the *Summa Theologica*. These concepts delineate the being of the subject, in the sense of *hypokeimenon*. In a clear parallel to Kant's distinction between receptivity and spontaneity, passive power relates to a primary act, or *forma*, that which is the aptitude or disposition of the subject. Active power relates to a secondary act or *actio*. Power is the spontaneous actualization of the subject. Leibniz rejects this distinction, critical of the concept of active power as *mere* power or potential, since it cannot provide an 'endogenous' account of the actuality of its possible act, remaining dependent upon an activation from a *deus ex obscura* of external stimulation.

Active force (**vis activa**), on the contrary, Leibniz explains, 'contains a certain act or *entelechy* (ἐντελέχεια), and is thus midway between the faculty of acting and the act itself . . .'[386] As 'in-between', active force, and as a rendition of the notion of intentionality, is a tending-toward, or a 'drive', a 'taking it upon oneself'. Drive, the 'basic metaphysical feature' of the *monadology*, is driven via an endogenous impulse; it is self-propelling. It is active in itself; this activity which can be sublimated, or 'set free', but its tension, as with a bow, cannot be stilled, as that would be its demise, its envelopment into the sleep of the bare monad. Drive can also be interpreted as *producere*, as a 'leading forth', to the extent that the subject is self-generated and self-maintained, as with the existential of Care. Heidegger describes the relations between monads in this context of 'aut-archy' as essentially negative: each monad asserts the limits to the 'striving' of all the others.

However, in such a situation of 'negativity', Heidegger asks, how are we to account for 'unity'? It is drive, which is alleged to confer 'unity', to pro-duce the 'unity' of the substance. Yet, if this is the case, and each monad is in a negative relation vis-a-vis all the others, how can Leibniz account for the 'unity' of our cosmos, of the uni-verse? Heidegger begins to chart an answer to the question of 'unity' through a consideration of the intimate self-relation of the monad, in the sense of an analogical relation between the 'I' and substance as aspects of the monad. Each monad, with respect to the analogy to the 'soul', is driven by self-activated changes, and it is this 'self-affection', as Leibniz asserted, that exhausts 'the entire sum of things'.[387] Indeed, the being of a self is projected as the measure of all things, as that which is given to things beyond sense and matter. What determines our be-ing is an 'understanding of being', which is 'itself' possible in that each of us is a being alongside other beings. Heidegger reminds us, however, that a concept of being has its 'proximate origin', not in beings, but in an 'undifferentiated presence of the world and ourselves'.[388] This 'presence', he continues, is 'primal transcendence', and it is here where there arises our concept of being. Heidegger states, 'An understanding of being belongs to the subject only insofar as the subject is something that transcends'.[389]

The interpretation of the being of beings, of substantiality, as that of monadic drive, as a primal transcendence from which arises an understanding of being, must allow an answer to the question of the origin of 'unity', or of how the 'structure of drive' is to give 'unity' to the being, to the ontic subject, and thus, to the logical subject. In other words, if the structure of drive can account for the 'unity' of a being, it will be compatible with truth as identity. In order to confer 'unity', drive must be simple. Leibniz claims, in the same letter quoted previously, that he inferred that there must be 'invisible unities in things' so as to avoid the 'absurd' idea, which would suggest that there be no original 'unity' in real things. Since this monadic drive is that invisible 'unity' which unifies, it must have a relation to that which is to be unified, as a manifold. With respect to the notion of identifying, and to self-affection, Heidegger states that this manifold must be engendered from amid the monad itself; it 'predelineates the possible manifold', which as an actively unifying drive itself, must be imbued with its own endogenous movement. Drive (**Drang**) presses (**bedrang**) the manifold, which as its own self, is a self-pressing, a self-driving, a self-surpassing, unifying in the heat of change and alteration, similar to Heraclitus' indication of *logos*, as the lightning bolt of Zeus that 'steers all things'.[390] The unifying substance, a prehensive (**vor-stellen**) drive, is the *a priori* 'origin and the mode of being of the changeable',[391] and thus anticipates a 'manifold'. Drive is an original 'unity' that reaches out towards, and surpasses that which it encircles in its *a priori* grasp.

Heidegger, while admitting that the ontological motive seems hidden from Leibniz, asserts that this 'structure of drive' is ek-static in its own incessant perceptive engagement with 'matters themselves'. It is gathered in an anticipatory 'one'. Drive abides within itself the original unity, an originary 'striving perception', for that which has been severed as thought and desire, *noesis* (νόησις) and *horesis* (ορεξις), of *perceptio* and *appetitio*. From amid this drive itself is born, in anticipation, the multiplicity of 'world'. Heidegger suggests that substance, drive, monad, entails internal change, and thus, a relationship to time vis-a-vis the temporal succession arising from the 'nature of the thing'. Or, to be blunt, Heidegger states, 'From drive itself arises time'; the multiplicity of 'world' is born from this striving, encircling 'tendency to transition' of drive, of substance, or, in still other words, of will. In this way, it is a striving perception which is to be, as substance, the 'unity' of the being. Such a 'unity' becomes possible with the uncovering of the origin of the manifold in drive itself. The origin of each individuated being from out of a striving, perceiving successiveness intimates the singularity, an 'identity' of each being which is thrown from amid the ecstasies of temporality, as each 'created' being is contained within horizons of the world.

Moreover, and returning to our earlier discussion of logic as self-regulation, in that this limit is constitutive for drive as such, there is amid its actualization an operative rule of limit or lawfulness. This rule is the anticipatory projection of 'unity', a 'look' from which drive comes to its own self-actualization. It arrives at itself in a return from its own surpassing of itself, in its advance 'view'. Amid this driving, surpassing, there opens a *topos* of self-openness, a clearing, amid which any being that is 'possessed' of drive may, come to 'know itself', to ap-perceive itself, to 'understand' its own being. An individuated 'identity' of existence is determined via its own unique 'event' of self-uncovering, a self-opening, its own actualizing actualization, amid its own singular existence.

One of the questions which arises from this interpretation of the 'unity' of the being is that of the 'unity' of the universe itself which is to composed of this Democritean 'chaos' of self-reflexive substances, each enclosed in its own temporal perspective, as its own 'representation' (temporalizing) of the world, enclosed in its 'windowless' self. Indeed, this is the question of the ontological sense of substance itself, a question Kant also raises, and which Heidegger points out in the sense of an ecstatic projection, that of anticipation, as a schema of pure, productive or transcendental imagination or temporality. Leibniz, in light of his notion of God as the supreme substance, answers the question of 'unity' and the interaction of the universe with his notion of a 'pre-established harmony', rooted in the will of God who is himself subject to the laws of reason. Each monad, in this view, is enclosed in itself as a 'living mirror' (**speculum**) of the cosmos. That which prevents a speculum from becoming the world as such is the finite character of drive, which discloses its limits in 'resistance', in the passive limit of receptivity which is wed to active, spontaneist tendencies of the monad (entelechy). In such a state of affairs, it is God himself who is the conduit for interaction of individual monads.

As we have been discussing, Heidegger has considered judgement, the 'state-ment' or *logos*, from a dual perspective, as a comportment of this be-ing of Dasein, the existence of 'my' self amid 'world', and, as an articulation, or expres-sion, of a determinative relation with beings. He states that this distinction is echoed in the 'bi-furcated intentionality' of statement itself as it joins and severs in its articulation by the figures of synthesis and diairesis. Amid this site of bi-furcation, 'unity' is possible, for Heidegger, in light of an intimate self-interpretation of one's existence, an analytic of Dasein which will uncover the relation between Dasein (and its comportment of judgement) and truth. Heidegger describes the character of such an analytic, as a 'metaphysics of truth'.[392] Truth, for Leibniz, concerns an 'inclusion' in the concept of the being. But, such inclusion concerns the identity, not only of the subject and predicate of a statement, but also, this 'identifying' of the being itself amid its substantial drive, self-organizing monadic be-ing. It is to the be-ing of this phenomenon that expression is to be dedicated.

> Thus the monadic structure of beings is the metaphysical foundation for the theory of judgment and for the identity theory of truth. Our dismantling of Leibniz's doctrine of judgment . . . is hereby accomplished.[393]

Heidegger finds a 'ground' for logic in metaphysics, of a relation of *logos* and being, in which the latter has priority. We are placed, once again, in the domain of the 'unsaid'.

Leibniz, as Heidegger provisionally concludes, cannot articulate a *logos*, expression, of concrete being in the world, existence as he remains locked in this turret of his 'windowless' substance, all relations mediated by God, refusing even Descartes' voyeuristic contemplation of passers by, wondering as to their 'actual' existence. Leibniz remains captivated by the traditional expression of substance which accounts for his own 'blindness' to a 'unity' amid 'contingency', rooted not in the authority of reason, but in the 'unity' of the temporal horizons of existence. In *Basic Problems*, Heidegger states that this windowless monad has no need of contact with that which it is not, as it possesses what it requires, what it is, amid its own specific degree of wakefulness. The 'world' can be 'conjured' up from within the self: it is projective, and is, 'in the end', oriented towards the 'outside'. In each monad, in conformity with its possibility there is represented a universe of all the other monads, the totality of all beings. Each monad already represents in its interior the whole of the world.[394] Heidegger, taking Leibniz to task, however, seeks to destroy the 'turret', the *mathesis universalis* of God and Reason, since the monad as drive is 'already outside, among other beings'.[395] The monadic self is an orientational 'perspective' of one's own 'phenomenal content', a pro-jection of world, an opening that allows things to enter into our freedom. In the next chapter, we will turn to the second half of *The Metaphysical Foundations of Logic* so as to more fully disclose the temporal, metaphysical grounding of *logos* in Being. The prevailing question will continue to be that of 'unity,' disclosed as the unifying unity of ecstatic temporality.

Chapter 11

The 'Unity' of Ecstatic Temporality

Things taken together are wholes and not wholes, something which is being brought together and brought apart, which is in tune and out of tune; out of all things there comes a unity, and out of a unity all things.[396]

Having uncovered in Leibniz's 'theory of judgment' a tacit relationship between *logos* and being, Heidegger seeks to specify the 'priority' or directionality between these modalities or comportments of existence. In other words, he seeks to disclose that which is responsible for the original 'unity' of these modalities of existence. Traditionally, the question of original 'unity' has been located in the problematic of the principle of reason (**principium rationis**), where principle refers to *arche* (ἀρχή), or, to ground. Yet, Heidegger asks the question of an originary 'ground' of 'reason' itself, disrupting the usual, facile, identification of the principle of reason as the fountain of original 'unity', and thus, raising the question of a relation of 'reason' and temporal existence.

Indeed, the question of the ground of 'reason' is revealed by Heidegger to be a variation of the *question of being*, or that which he calls the principle of 'rather than' (**potius quam**), as in the question 'why is there something, rather than nothing?' The *why* of a factic something indicates the question of the relationship between possibility and actuality, of the factual existence of that which is, with respect to the question of its ground of being-there. A being that is *actual* is one that has come into being, Heidegger suggests, due to its fertile 'essentiality, possibility'.[397] That which stands between possibility and actuality is 'rather', a moment of 'preference', of freedom, a preference which decides that which is to be and that which is not to be. In this light, the question of being becomes that of freedom as the ground of being. Heidegger suggests, as he had done in *Basic Problems*, that the question of ground intimates the ancient conception of the good, of *agathon* (ἀγαθόν), a Sun, which, like time, lies 'beyond being' as the 'ground' of being. He comments that, as with Plato's *Myth of the Cave*, the good is the light that is necessary for understanding beings *as such*, as it is an *a priori* source of orientation. It is that which is 'beyond being', and gives 'unity' to finite existence.

In the following, I will lay out Heidegger's excavation of the principle of ground to the jointure between *logos* and being. Beginning with the primacy of the 'principle of reason' in Leibniz, Heidegger will disclose through a genealogy (recollection from historicity), of the concept of ground in its various historical interpretations, the authentic ground of freedom which is suppressed in the doctrine of reason. The temporal ground of the principle of 'reason' will be unearthed, which will reveal the pure *sensible* character of finite 'reason', of an understanding of being amidst being-in-the-world. It will be uncovered furthermore that an original 'unity' of existence is due to ecstatic temporality and not, contrary to Henrich, to traditional 'reason' or subjectivity.

The principle of ground

A radical phenomenology of existence finds its point of departure in the pre-understanding of existence in the sense of being-in-the-world. It is amid such an original orientation that our first senses of ground arise. Ground is understood, for Heidegger, with respect to factical comportments of *logos* and being. In this way, ground has a pre-sense of 'cause' *aitia* (αἰτία), as with the 'cause' as a ground or proof of an argument. This is an understanding arising amid the factical 'event' of everyday speech. Yet, in this pre-understanding of being, it is *techne* (τέχνη), or, the everyday production of beings by means of beings, that gives to us a pre-understanding of 'causation'. It is amid this play of *logos* and being that beings 'come to presense' in the indicative or truth-telling character of *logos* (λόγος) as *episteme* (ἐπιστήμη), as the disclosure of the what-it-is (τί ἐστιν) in an *eide* (ειδη), a *look*.

   Heidegger reminds us that the question of the *a priori* grounds of true statements, as the question of ground, cannot be simply abstracted from factical realities of everyday 'experience' and set up as a normative standard 'once and for all'. In the everyday world, even if we could conceive of an ideal coincidence of the grounds of truth and those of certainty, or if certainty could 'wait' for an explication of the grounds of truth, we still would not have uncovered an *a priori* ground, but a merely normative rendering of the question of ground, entangling one into the role of an inspectorate of particular grounds, enclosed merely in the realm of real predication. Heidegger states that his excavation is not concerned with the grounds of specific truths, but with the relation of the essence of truth to 'what we rightly call "ground"'.[398] It is this explication which, he contends, will provide support for the derivative problem of the ground of true statements, which is 'necessary in some respects'.[399] The *destruktion* will set out from a historical determination of the 'essence of truth' from which may be excavated a more originary ground. As we have already encountered, the essence of truth is located, for Leibniz, in the judgement or proposition (λόγος). In this way, 'truth' is the validity of a 'combination of representations', *synthesis noematon* (σύνθεσις νοημάτων), as an adequacy of the 'belonging together' of a subject and predicate. Heidegger claims, however, that this portrayal of truth cannot touch truth in its essence in that statements or propositions are dependent upon that which decides what is to belong together, or the 'object'. A statement is indeterminate without its correspondence, as with a representation, *noema* (νόημα) to the 'thing itself', *pragma* (πραγμα). It is the being that 'measures' the truth or falsity of a statement in the sense of adequation *omoiosis* (ομοίωσις). Heidegger, rejecting an 'epistemological' interpretation of this correspondence, seeks a beginning amid pre-theoretical/practical comportments of being. He asks:

> How does the proposition present itself to us, when we comport ourselves prior to all theories of judgment, prior to all philosophical questions about propositions?[100]

Heidegger asks to what do we attend when we hear or read a statement. Echoing his comments on the 'Greek' sense of the *a priori* in his *History of the Concept of Time*, he asserts that a statement directly indicates the being itself as the 'about which' of the former, as for instance, when we look towards a cactus that is indicated, it is an integral being, there for our apprehension. The 'right way' of investigation is thus an emphasis upon the 'primary givenness' of the thing (**Gegenstand**) which sets the limit and domain for the inquiry. It is with these criteria that we are to grasp the essence of the statement, not as some self-referential 'pure' logic, but simply as a 'statement about something'. Implied in such criteria is that a statement is dependent upon the primary givenness of beings and upon a comportment with/among these beings. In other words, the statement about beings exists within and is dependent upon the horizon of a prerequisite understanding of beings. In this way, the contention that truth finds its 'locus' in statement is misleading. Indeed, a statement, as rooted in being-in-the-world, is merely a 'sign' of this deeper, more original truth of primary givenness. Such a being amid beings and the *a-lethic* truth that it entails, is rooted in *existence*. It is to this extent that the be-ing of existence, as being-open, is distinct from 'nature' conceived as a manifold of 'objects'. It is here where we can fathom the intimation of an expression beyond real, theoretical predication, in the self-expression of indigenous existence.

Existence as a being-by beings-in-the-world, is essentially disclosive, and thus, indicates the original sense of being truly expressed in the Ancient Greek *a-letheuein* (ά–ληθεύειν). Truth, in this sense, has a privative character, the negative resides in it. Being true is dis-closure, a wresting free that belongs to our existence. The truth of the proposition is derivative, it is only one sort of truth, intimating that there is a primary perversity of truths amid being-in-the-world, recalling Brentano's early emphasis, following Aristotle, upon the many ways in which being is said. Not only is there 'plurivocity',[101] one that includes a possibility of silence, but there can also be distinguished original and derivative senses of truth. In this way, Heidegger argues, that an original meaning of 'ground' must take its meaning from somewhere else, 'beyond' judgement.

Intentionality and the problem of transcendence

The original sense of truth, a stranger to the restricted truth of judgement, is characterized by a dis-closure of beings in light of a primary comportment with beings, of a being-by (**bei**) beings. In this sense, original truth is concerned with the pre-theoretical and pre-practical sense of the standing-against (**gegenstand**) of beings. Heidegger's provisional interest is a 'natural' relation to beings, bracketing the logic of the problem of a subject-object relation, which, as a 'pseudo-problem', vanishes.

> Intentionality is indeed related to the beings themselves and, in this sense, is an ontic transcending comportment, but it does not primordially constitute this relating-to but is founded in a being-by beings. This being-by is, in its intrinsic possibility, in turn grounded in existence.[102]

Intentionality, conceived as an 'ontic' transcendence, is founded upon the primordial transcendence of being-in-the-world. This transcendence intimates the 'metaphysical neutrality' of Dasein conceived as 'the origin', as the radix, of its own thrown 'dissemination' (**zerstreuung**) as the world. Neutrality, as Heidegger explains, indicates the 'metaphysical isolation' of freedom peculiar to our own existence. That which he seeks, amid an echo of the work of Dilthey, is a conceptual expression attuned to Dasein, the be-ing of the being that we are. Such an expression, Heidegger states, is only possible in the act, amid the free projection or an intimate involvement, of one who seeks to understand the being of the being it is , that is, it must be an indigenous expression of a singular, temporal sense of being, and not of the wholesale re-branding of 'known knowledge'. He insists on this *special* character of finite understanding – it is 'on the way', makeshift, with no legislative intent or pretensions.

In a further elaboration of the discussion in Chapter 4 concerning the temporal 'ground' of understanding, Heidegger asks: 'But why is time connected with the understanding of being? This is not obvious'.[103] He points out that an 'identification' of thought (νοειν) and being (ειναι) by Parmenides, is not a merely subjective absorption of being, but, instead, ειναι is 'not yet clearly differentiated from the ov'.[104] In this way, Heidegger indicates a relation of 'subjectivity' or finite knowing, which, in reference to the *Theaetetus* of Plato, he also calls soul (ψυχή), to being (ov). Such a relationship is a striving for being by the soul, and it thus suggests a differing and, for Heidegger, a more original sense of subjectivity, or that which he repeatedly refers to as the 'subjectivity of the subject'. But, this brings us back to our question of the relation of time to an understanding of being. Indeed, the relation is already indicated in the reference to finite knowing, one that is not able to grasp being in its totality, or in other words, it is a finite understanding of being. That which is at issue for Heidegger is not whether such an understanding is 'in time' as is the case with an 'intra-temporal' being, but if such an understanding is comprehensible 'in relation to time'.[105] The question is of a 'relation' of 'being and time'. For Aristotle, being as such, being-ness (ουσια), or being-as-being (ον η ον) is determined as both modes of existence and essence. The former is that which is on hand, the that-being or primary substance, while the latter is what-being or what-contents, that which we have already seen as the Latin, *possibilitas*, that which something 'is' whether or not it 'actually' exists in the sense of *actualitas*, or, secondary substance. On the one hand, both modes have a relation to time in the sense of a constant duration of a being that always is, *aei on* (αει ον), in which existence refers to its constant being and where essence, as an idea, determines a being, in advance, as a being. Heidegger reminds us of the pre-philosophical meaning of ουσια as possessions, property, which is an everyday expression of the *what* that is present 'in every now'. It entails, in this way, a temporal determination with respect to the meaning of 'presentness' (**Anwesenheit**).

Yet, these references remain problematic for Heidegger since they are ontic determinations of being, and thus are ambiguous in that they fail to make necessary distinctions with respect to not only the meaning of being, but also regarding an original sense of temporality. As we have suggested, being is that which is 'earlier' or prior to beings. In other words, it is that which determines beings 'in advance' as the *a priori*. Heidegger, in an echo of his discussion in the *History*, lays out an original sense of an *a priori* as that which is prior 'by nature' distinct from what is prior 'by knowledge', *proteron phusei* (πρότερον φύσει) and *proteron gnosei* (πρότερον γνώσει), respectively. Priority by nature is 'essentially prior', or, in the language of *Being and Time*, it is the 'transcendens pure and simple'. Priority by knowledge is that which is 'prior with regard to us', it concerns the order of our ontic conceptualization. For Heidegger, the 'prior to' indicates a temporal orientation, not in the sense of a 'common time', for, in the order of conceptualization, Heidegger states, 'being' is the 'last of all'. Being as 'prior by nature' is neither a being nor is it a logical priority, but is the 'earlier' as such, the *a priori* horizon.

As an attempt to begin to clarify the obscurity of the priority of being, he invokes Plato's doctrine of 'recollection', *anamnesis* (ανάμνησις) in the *Phaedrus* and *Meno*, as a primitive self-interpretation of Dasein, a remembrance narrated in the myth of the 'transmigration of souls'. This myth, as an indicative nexus, or complex symbola, gives an account of the possibility of an 'earlier' as such and is reputed to be the 'ground' of the *a priori* in Pythagorean and Platonic philosophies. Heidegger lets this myth to speak us, in our own idiom:

Being is what we recall, what we accept as something we immediately understand as such, what is always already given to us; being is never alien but always familiar, 'ours'.[106]

In this interpretation, recollection refers to our pre-understanding of being, an understanding, as we will recall from our previous discussions, which becomes covered over in standardized normality. This sense of being remains with us, but falls into the background amid our captivation in beings. It is *metaphysical* recollection, Heidegger declares, that allows us to come back to an original understanding of be-ing. It is in this sense, among many differing senses, that being has an original relation with time, as an intimacy sheltered in pre-philosophical expression.

The intimacy of being and time, and thus, of being and transcendence, and being and truth, incites an attempt to express the relation of being and time with respect to the self-interpretation of the 'analytic of dasein'. Heidegger, seeking to radicalize the Platonic question of being, states that questioning after being requires an original interpretation of time, of original, ecstatic temporality, and its relation with the being of human Dasein. The relation of Dasein and time is exposed in the light of the understanding of being which is an indigenous aspect of existence. In that being is linked to the *a priori*, it is also connected to temporality, and thus, to Dasein, which is 'in the truth'. Being is the prius of beings – each is its event. There are not only diverse regional phenomenologies of existence, of history, of artworks, of nature, but also the 'unity' of these myriad 'ways of being' – a 'unity' which is misunderstood as *being in general*. The distinction between regions and their *unity* intimates the 'ontological difference' which lays out the distinction of being and beings. 'Being' is not a being, not for the hand as such (**vorhanden**) or to the hand in the concrete situation (**Zuhanden**). Being is 'there' where the question asserts itself, where being becomes an issue for finite Dasein as it asks 'what is being, why is being?' In its traditional articulation, 'being' pertains to an essence and existence of a being, of its *what*, its content, and of its *how*, its actuality. Heidegger, taking a step back from essence and actuality, opens up a phenomenal place, where in the 'essence of existence there is transcendence', indicated as the Open, clearing (**Lichtung**) which gives (**es gibt**) light to beings. This 'no-thing' of being happens before, prior to beings, as a horizon for beings, it projects a light through which beings can come into truth.

From transcendence to being-in-the-world

As we have seen, Heidegger situates the dimension for unfolding the 'problem of ground' in the question of being. This latter question moreover requires that the 'one', the questioner, enters into self-questioning, into self-interpretation. Such an 'analytic of dasein' prepares a *topos* from upon which the being-question (**seinsfrage**) may be asked. Traditionally, transcendence has been interpreted as the Latin *transcendere*, to step over, to go beyond – it is action, relation and a limit to be transgressed ('in-between'). Heidegger brackets traditional ideas of transcendence, as each corresponds either to immanence or contingency, and thus, there is either a mythology of consciousness, or one of an Absolute. Heidegger dissolves this Janus-faced portrayal of transcendence by pointing out that over against the epistemological question of beings which relate to one another, subject and object, there is an 'eminent' existence which not only transcends each of these beings, but also grounds them and their epistemic relation. He describes this entanglement as a rift between ecstatic transcendence, as being-in-the-world[407] and, that which overwhelms existence as such, *intuitus originarius*. Significantly, he states that it is from the entanglement of this confusion that Kant, in his own way, tried to escape, although unsuccessfully.

Heidegger seeks to throw off the language games, received metaphors, images – spacings – which sustain the traditional discourse of 'consciousness' and the authority of Reason as the pillars of the notion of transcendence. He states: 'Transcendence is rather the primordial constitution of the *subjectivity* of a subject'.[408] The subject is 'transcendence', is *transcending*, and nothing else besides, always ecstatically open, 'Dasein is itself the passage across'.[409] It is 'neutral'. Beings become accessible to Dasein since they have already been surpassed, just as with the myth of pre-existence in Plato. However, contrary to the latter, the beings gain access to the open-not, for Heidegger, as an absolute mysticism, but amid this playspace of *a-lethea*. The opening of Dasein becomes a protracted uncovering, 'discoverability' of that to which traditional notions have already *covered over*. Being-in-the-world as the transcending of Dasein does not refer to 'extant' actuality, but to a way of being, finite transcending. In this light, 'world' takes on a different meaning from that of the Kantian 'Idea' of the world as the totality of extant things, or as 'nature', and of the meaning of being as positedness, of things subject to 'consciousness'. Existence is, on the contrary, an ecstatic openness, 'outside' of the hegemony of factual actuality and real predication. As the *topos* of original truth, as world, it is the 'ground' of statement (λογος).

The phenomena of world and freedom

Heidegger enters into another historical recollection, in this instance, remembering the meaning of world, κόσμος, from Parmenides and Heraclitus to Kant. Consistent with his point of departure in the Presocratics, he emphasizes that κόσμος does not refer to the totality of extant beings, the existentiell, ontic domain of entities, as he claims is the case with 'Kant'. On the contrary, Heidegger states:

> κόσμος refers rather to 'condition' [Zustand]; κόσμος is the term for the mode of being [Weise zu sein] not for beings themselves. κόσμος ουτος means the particular condition of beings, this world of beings in contradistinction to another. Beings themselves remain the same . . . their world can

differ; or one can hold the view that the world of beings always remains the same.[110]

The *early* Greek meaning of Kosmos, world is, for Heidegger, active, as it means 'to world', in the sense of an original condition for the possibility, a *topos* for the factical domain of beings. World is a 'worlding' which is an original 'unifying' modality, the 'how' of a 'totality'. It is understanding, for Heidegger, and not the eyes, which comprehends this unity which lays out the context of possibility for the coherence of all beings. As with the notion of an ontological 'unity', which serves as the condition for the severance of the stems, Heidegger points out world as the 'basic condition', as the prior, and reminiscent of Kant's notion of pure formal intuitions in the First *Critique*, from which there is a subsequent 'partitioning'. For Heraclitus, this oneness and unity of the world, 'belong to the awake, but each of the sleeping turns to his own world'.[411] Heidegger contends, however, that the phenomena of 'world' pertains to the mode of the being of beings, to the 'unity' of being, to the movement or temporality of being, and it is oriented to the be-ing of Dasein, whether 'awake' or 'asleep'.

Shifting perspective, Heidegger turns to the concept of 'world' for the early Christians in Paul's 'Letter to the Galatians', a text with which he was concerned in the early 1920s (before, that is, he was 'shattered' by Nietzsche). In this interpretation, world (κόσμος οντος) maintains the sense of condition, a mode of being, but evokes, specifically for the early Christian cult, a specific way of being, *ethos*, or comportment towards beings of a community, which in its own self-interpretation, is 'forsaken of god'. Amid its temporal situation, its world is oriented to the anticipation of its future, to the age to come, its final situation (εσχατον). Two centuries later, Augustine, a Christian Neoplatonist at the close of the Roman Age, begins to transform the meaning of 'world' (**mundus**) from its association with the totality of beings of nature, to the fallen domain of the flesh, of a god-forsaken human existence, in absolute contradistinction to the radical eternality of the divine. Heidegger mentions briefly that this notion of 'world' loses the 'anxiety' of its earlier eschatological expression with Thomas Aquinas, and becomes the secular in distinction to the spiritual.

Despite his captivation to the description of world by Baumgarten as an 'aesthetical' collection of worldly things, of museum pieces, dead objects, Heidegger contends here that Kant had projected, but immediately suppressed, a transcendental, or ontological, concept of world, in which he carved out some space 'beyond' the ontical conception of world as a 'series of finite actualities'.[412] 'World', in this sense, is a totality of the possible, an opening of transcendental understanding, distinct from the extant, actual world of physical understanding of a 'composite' world. Heidegger elaborates upon this suggestion by following the distinction between mathematical and dynamical principles that distinguish the world from nature. He contends that the mathematical, or intuitive, principles concern the essence or the *what-being*, of a natural thing (ontological-essential), while the dynamical principles refer to actuality, or *how-being*, as the *modus exis-tendi* of the natural thing (ontological-existential). Kant writes in his *Critique of Pure Reason*, however, that each of these principles of pure understanding, are treated only 'in their relation to inner sense' (A162, B201–02), or, time, and each, world and nature (φύσις) refers to a self-same totality, or reason, but from the differing aspects of essence and existence. Since such a notion would orient itself within the domain of real and actual predication, of 'essence' and 'exist-ence', he rejects this conception of 'world' in favour of the temporal projection of an *a priori* horizon of world, of an act, or, event, that surpasses beings with the opening of a horizon for an originary emergence of intra-worldly beings. It is 'world' that is the 'towards which' of projection for an understanding of being, but, whose characters are the projections of ecstatic temporality, or, historicity.

Dasein, in its finite transcendence, passes over its own self, and in surpassing itself, an 'abyss' is opened which it, in each case, is for itself. An abyss can be covered over and obscured, only because the 'abyss of being-a-self is opened up by and in transcendence'.[413] Yet, what is this *world*, the transcending towards which, opens up an abyss for this being-a-self? Heidegger rejects the idealistic solution of Plato, which, despite its gestures towards an 'otherness', is merely an imitation of the appearances of the natural world and is thus a detour from a situation of original truth. He states, 'The connection between ideas and looking, θεωρία, intuitus, (referred to already in the word ιδέα), is essential, since the source of the doctrine of ideas is expressed in it.'[414] It is this vision-dominated interpretation of transcendence which eventually leads, Heidegger contends, to the epistemological rift between subject and object. The twin of this theoretical interpretation is located in an aesthetic intuition, which is alleged to grasp the beautiful in disinterestness. None of these 'signs', *theoria* (θεωρία), *aesthesis* (αεσθησις) and *praxis* (πραξις), can disclose the phenome-non of transcendence, in that *existence* is a basic phenomenon of an *a priori* 'unity'.

> The central task in the ontology of Dasein is to go back behind these divisions into comportments to find their common root, a task that need not, of course, be easy.[415]

The question of existence concerns the relation of transcendence, being-in-the-world, and the 'problem of freedom'. Heidegger frames the question of 'unity' with respect to the distinction between spontaneity and receptivity. What is to be disclosed is transcendence, a clearing that is prior to these stems; the 'radix' of an *a priori* unifying movement, of original temporality. Heidegger returns to Plato's 'idea of the good', τος άγαθον, for a clarification of the sense or mean-ing of a 'beyond' of transcendence, of this place which is no place, of a *topos* that is nothing and nowhere. He interprets this 'otherness' as a root, 'before'

the thrown dissemination of beings, in which these beings trace their orientation of existence. There will be, for Heidegger, however, no Platonic realm,
however, of pure forms, conceived as other-worldly ideas, but instead, the otherness will refer to the transcendence of Dasein. This more original interpretation
of the idea of the good, Heidegger states, refers to oυ ενεκα, to the 'for-the-
sake-of-which'.[416] This latter is 'world', disclosed as that which excels (ἐπέκεινα)
ideas and beings, but, at the same time, organizes these in its 'communality'
koinonia (κοινωνία). World is a 'purposiveness', which is '. . . only possible
where there is a willing'.[417] Transcendence, being-in-the-world, and freedom
are thus the same.

The free responsibility of the self-questioning self

The question of the 'for-the-sake-of-which' is concern for the existence of the
one who questions, and arises from the concerns of a questioner, who can alone
'pose the question in its real sense and answer it'.[418] The truth of being concerns the one who exists in question; the 'questioner's situation is included in
the question'.[419] Heidegger refers to this self-concern of the questioner as a
'metaphysical egocity of dasein as such'.[420] Careful to dissociate this concept
from some ontic or 'existentiell ethical egoism', the metaphysical sense of
'egoicity' [**Egoitat**] refers to that which we have referred to as the 'metaphysical
neutrality', or 'isolation', of Dasein, which as freedom, is the abyssal ground of
extant beings. In other words, in order for Dasein to be for other Daseins, or,
for intra-worldly beings, it must cultivate its own being. It must be concerned
for its being-as-a-self and of its possibility, with its 'mineness and selfhood as
such'.[421] Selfhood is in this way the 'for-the-sake-of', the purposiveness of this
existence of dasein as being-in-the-world. Existing is a basic being towards oneself, as an intimate self-existence, and as free, it is, at the same time, a choosing
of oneself from out of its possibility, from an understanding of its being, even if
this understanding 'makes sense' only to the questioner. Heidegger warns that
this 'singularity' is not meant to convey a solipsistic contraction of an ontic self
into its own ego, if that were even possible. That which is, is already a be-ing
there with others and things, these are there amid one's finite horizons of
possibility. However, in its freedom, Dasein attempts to choose itself out of its
possibility, for the sake of itself. Singular existence implies a commitment of
a self to this existence.

The for-the-sake-of is in this light uncovered as the structure of the world,
as the *towards-which* of transcendence as being-in-the-world. The *for-the-sake-of* as
the *towards-which* of transcendence is the destination of freedom, of a willing.
But, these are not distinct things, or stems; freedom is the origin of the for-the-
sake-of, in the sense of being 'at one' with this self-purpose. Freedom, echoing
Bakūnin, is its own 'purpose', its own 'means'.[422] Yet, contrary to Bakūnin's
traditional concept of freedom as a spontaneist, and hence subjectivist act,

Heidegger emphasizes this situatedness of freedom in the being of the self amid Dasein, a collective, temporal existence from amid which my self emerges into the light, coming to stand. In other words, he seeks to disclose receptive dimensions of finite transcendence, and horizons of freedom that emerge in the open of our own original temporality.

The for-the-sake-of is a self-understanding which a free Dasein gives to itself. Freedom, as an understanding of being, is the projection of a purposiveness of Dasein. World as a for-the-sake-of, is, Heidegger states, the 'primordial commitment' [**Bildung**] of a free imaginative Dasein. In its freedom, Dasein is responsible to itself, it gives itself a possibility of commitment. Its commitments are its 'world'. Dasein in its situated freedom projects 'world', it holds itself in this world as that which is binding to itself, a counter hold that preserves a self-chosen commitment. It gives to itself horizons for the entry of beings. Dasein, Heidegger states, is excessive, insatiable for beings, for possibilities, 'it' surpasses these amid its finite transcendence. But, this surpassing of beings is a free projection of a world that opens up the limit horizons within which beings can be encountered amid its 'world'. Dasein surpasses beings and its own self. It is ecstatically open, 'outside' itself, it is thus being-in-a-world, open to beings. Amid this openness of self-transcending, a 'temporalization of temporality', beings enter 'into world'.[423] World-entry has the characteristic of happening, of history [**Geschichte**]. World-entry happens when transcendence happens, that is, when historical Dasein exists. Only then is being in the world existent. And only when the latter is existent, have extant things already entered world, that is, become intraworldly.[424]

World-entry is not a causative principle of the production of beings, objects, but refers to the phenomenological awareness of Dasein amid the original possibilities of its encounter with beings. It always encounters beings, which would exist with or without Dasein. Yet, these beings enter into this world of Dasein in the sense of being there in the free projection of binding commitments. Beings come to be in our world, there, figure in our world, they become framed in our world, amid our free projection of self-responsibility. The world is a 'nothing' – not a 'thing' in the world, but instead a projection of temporal commitments amid ecstatic temporality. However, as we will ascertain, these commitments can be unknotted via time, they are makeshift.

The 'unity' of ecstatic temporality

In this lived 'movement' of Dasein, the latter, be-ing amid transcendence, is 'outside', 'beyond' itself. Be-ing 'outside', in its movement, it exists towards a destination. In its movement, this projection clears a *topos* of be-ing upon which Dasein is 'suspended' as in an array of possibilities towards which it moves. As free, Dasein chooses its possibilities, it orients itself to these chosen possibilities as a makeshift array of binding commitments.

Dasein, amid its ecstatic self-projection, holds before itself its world, it holds itself 'beyond', in the 'no-thing'. This 'beyond' of projection is not a mere nothingness, *nihil negativum*, but as we will recall from Part 2, an *upon-which* that is a horizon of ecstatic, original temporality. In an attempt to uncover the ecstatic character of original time, Heidegger, not content to merely introduce a 'new' theory of time, examines our lived discourse of time, or, our utterances with respect to time, the names we give to time, such as 'now', 'then' and 'at that time', expressions which 'speak of time'.

That which is uttered in these expressions, such as *now – a door slams*, is a relation between the temporal expression 'now' and an event. The expression 'now' is not an isolated temporal thing, nor is 'it' to be compared to an event determined or regimented via an external measure of time. Instead, the 'now' is an indicator of an event, it points to an event, and thus, as Heidegger states, it has a 'forward indexical function'.[425] In an emphasis of the dependency of the temporal expression 'now' upon an event, he states that the event always happens prior to this expression of its occurrence 'in the now'. Heidegger turns to the 'then', which has an orientation to 'futurity', towards the 'not yet', that which is expected or anticipated. Unlike the clarity of the now, the 'then' stretches out towards a peculiar 'nothingness' in a projection upon that which remains 'indefinite and indeterminable'.[426] An expectation is indicated and uttered by the expression, 'then'. Expectation, as a mode of existence, is a projection which carries with it, in a similar way to the double projections of the ecstasis of original temporality, its own 'then', its own target for this mode of expectation. This is not a 'then-thing' outside to which it would be related as to an object, but is an indication of a directionality or tendency of an indigenous mode of temporalization.

Expectancy, as an intentional comportment 'toward the futural', is a being ahead of oneself, but as a comportment of self-relation, through which the being comes towards oneself in the fulfilment of an expectancy, as a possibility for being of the self. Heidegger states:

> Expecting one's own capability-for-being as mine, I have also come toward myself already and precisely through expecting. This approaching oneself in advance, from one's possibility, is the primary ecstatic concept of the future.[427]

Expectancy, Heidegger states, is 'ecstatic', 'stepping out itself' (εκστασις), and 'to some extent a raptus [rapture]'.[428] The original projection of transcendence is the *a priori* 'ground', an opening that clears the playspace for a specific modulation of expectancy, of hoping, fearing. These are concretely possible in the midst of the thrown dissemination of horizonal ecstases of a temporalization of an original temporality. Heidegger seeks to be unambiguous: this temporality of Dasein has nothing linear, common about it. The 'excessive' stepping out of temporal Dasein is disseminated amid an 'open path made way by the raptus of temporality itself'.[429] Dasein temporalizes its past and present 'only from out of and in the future'.[430] Neither the present, nor this having-been, is in this way a fixed quantity, dead artifact, contained in itself, inaccessible – quarantined from this 'living' in a museum. Each is 'on the way', an existence always being temporalized with respect to the thrown dissemination of existence, and of the self-interpretation of this existence. For Heidegger, the making present of the moment, of a 'now' arises amid an 'ecstatic unity' of a futurity and having been, suspended in-between the ecstases, which 'are themselves ecstatic' amid a self-unifying transcending of 'free ecstatic momentum'.[431]

     Temporality as a 'free oscillation', Heidegger continues, finds its 'unity' in an 'ecstatic unitary oscillation'.[432] These horizons are projected with each ecstasis and owe their 'unity' to that of ecstatic temporality. The horizons (ὁρίζειν) delimit this *topos* of disclosure, the 'Da' that clears this playspace, this place, of existence. A future as a possibility for existence, while it intimates 'positive' possibilities, at the same time, lays out limits for being, horizons of finitude, as with our *being-towards-death*. Each ecstasis projects a horizon of its own unifying self-enclosure, within which there can then enter a specific expectation. Ecstatic self-transcendence is not a transgression of every barrier, but a surpassing which projects the horizon for its own space for being, 'eventual' object related-ness, for experience. Yet, since it gives no determinate being, this horizon is, as is being, *nowhere*. As a projection of ecstatic temporality, it 'temporalizes itself' amid the oscillation of ecstatic temporality. Related to our discussion in Chapter 9, and as a further clarification of that discussion, a horizonal-schema is an 'ecstema' of this 'ecstematic unity of the horizon of temporality', temporal situation for the 'possibility of *world*'.[433] Transcendence is possible in an 'ecstatic momentum'.[434] The ecstematic, the implicit unity of the ecstasies, temporalizes itself, oscillating as a worlding [**Welten**]. World-entry, Heidegger concludes, happens when and only if ecstatic oscillation temporalizes itself.[435]

     Heidegger states that 'world entry' is primal history and is related to the mythic as a primitive self-interpretation of Dasein, of what we have repeatedly referred to as the pre-understanding, or the *prius*, or, of this peculiar *a priori* of a radical phenomenology. In other words, 'myth' can be interpreted as an original projection through which a self-understanding will emerge, an under-standing of being, itself a singular excession of transcendence. Heidegger returns to Leibniz's 'myth' of a 'windowless monad'. This 'image' is questionable, he states, in that it entails in its 'aesthetics' an adherence to the Cartesian *ego cogito*, in its subjection of substance, in its 'containment' within itself of every possible exterior. Indeed, for Heidegger, a windowless monad could only make sense in a world of which there is no 'inside and out', where there is this self-opening of an 'ecstatic happening of world-entry'.[436]

     This 'unity' of ecstatic temporality, the 'ground' of a 'world', is a for-the-sake-of, purposiveness of one's binding commitments, a promise, projected 'into the future'. One comes back towards oneself from this projection in fulfilment of a promise, commitment, or, in its betrayal or forgetting. Projection is this temporalization of temporality amid finite existence 'where' beings enter into a 'world' composed only of shifting projections, oscillations of temporality. Self-understanding emerges amid 'primal history' as intimate self-interpretations of 'philosophy', which, for Heidegger, is the expression of the selfhood of Dasein. The essence of ground lies in that which gives 'unity' to a 'primal' history of Dasein, that which bestows this 'unity', meaning of existence, its purpose.

Heidegger states, that his seemingly novel lexicon, his 'jargon' of ecstasy, rapture, thrownness, etc., is necessary in that this 'unity' he seeks emerges amid these projections of ecstatic temporality, and thus, emerges *in and from freedom.* Freedom cultivates itself as an intimate self-understanding and self-expression of its existence, meaningful only in its relation to its own lived temporality. Its primal history, 'temporalized' out of its own projections of a for-the-sake-of, of meaning, purpose, is grounded upon a freedom which 'is' and understands itself as the origin of responsibility. Because the for-the-sake-of is the recoiling for-the-sake-of-itself, freedom is, out of necessity, ultimately the *ground of ground.*[137] The 'good' is a 'nothing' projected upon a horizon of this future, as an aspiration for a possible future.

Chapter 12

The Riddle of Fallenness, the Building Site of Care and Temporality

A philosophy is never a house; it is a construction site. But its incompletion is not that of science. Science draws up a multitude of finished parts and only its whole presents empty spaces, whereas in our striving for cohesiveness, incompletion is not restricted to the lacunae of thought . . . at each point, there is the impossibility of the final state.[138]

I will turn now to an exploration of the intimate be-ing of Dasein, which is indicated in *Being and Time* as Care. Although this latter work has haunted the background of the preceding chapters, I have made an explicit attempt to provide an interpretation of Heidegger's 1920s phenomenology that would treat *Being and Time, Kant and the Problem of Metaphysics* and the lecture courses as threads within the greater tapestry of the 'Sein und Zeit' project. These differing expressions are necessary in that, for Heidegger, being has a priority over *logos*. And since the meaning of being is temporality, the expression of the truth of being will be compelled to ceaselessly shift in order to remain in the truth. The makeshift character of this thinking intimates the Winter lecture course of 1942–43 on Parmenides and Heraclitus in which Heidegger speaks of an intimate sense of translation in an appropriation and use of language. As there is no *essential* language in itself, each articulation becomes translation, amid a *topos* where meaning is also in a state of flux. The character of 'first philosophy', as that which must always be started anew, points to the significance of the lecture courses and the Kantbook as novel expressions that are guided by the movement of the phenomenon. *Being and Time* was also one of these novel expressions.

As with Chapters 3 and 4, the next two chapters will be laid out as prospective and retrocursive projections, with an exposition of the temporal characters of being and the horizonal-ecstases of original temporality, respectively. In this light, the existential of Care is a prospective expression of the be-ing of Dasein. Yet, it must be remembered that its meaning is temporality, a meaning which is appropriate in that Care had only emerged into expression, retrocursively, through a temporal 'event'. In this manner, the circular 'relatedness backward and forward'[139] is an index of the temporal 'stretching out' of this *topos* for phenomena 'there'. This circularity reveals the temporalization of Care, which is disclosed in a radical singularization of finite self-knowing, without evasion, in anticipatory resoluteness. In keeping with the indication of a *topos as a pathway*, I will evoke the incompleteness and ambiguity of Care with respect to the meaning of the 'who', this question of *which* 'interpretedness' has the strongest hold on the sense of the being of the self. Care harbours within itself an ambivalence and fragility with respect to the 'truth' of oneself; it intimates the radically temporal character of existence and the inevitability of 'events' of self-disclosure and transformation amid one's singular possibility of existence.

Taking our point of departure in a disposition of anxiety and of individuation, the self travels ever nearer towards its being, towards attunement and openness with respect to its personal existence. The be-ing of the personal and of Care, projected amid the disclosure of the 'matter themselves', is pro-voked by a sense of uncanniness, of answering a 'call' from the singular self to the anonymous self in its untruth. We will consider the resonance of the fragment of Anaximander with respect to the meaning of temporal 'guilt', an indication which discloses the radical temporality of existence, or, as Schürmann has pointed out, tragic existence. In the following pages, aware of the peculiarity of the 'methodology' and its formally indicative significance, I will first sketch out a topography of thrown projection, of the 'there' in *Being and Time*. I will next lay out Care as the Being of Dasein, disclosed in the existentiell disposition of Anxiety. Finally, I will intimate the radical incompleteness of Care, if considered in itself away from its event of disclosure, of its 'ambiguity' and 'falling' with respect to its own ceaseless departure from its event of emergence.

The *topos* of 'thrown projection': The 'there'

Heidegger lays out in his *Being and Time* a complex indication of the 'there' of existence. This serves as a specification of the indication of being-in-the-world as the disclosedness of Dasein. There are many variations of the constitutive aspects of this *topos*, but it repeatedly hovers around understanding, disposition (**Befindlichkeit**),[110] falling, and discourse. Each of these 'structures' is a temporal modality of the 'ground constitution' of temporal existence, comportments of Da-sein which disclose existence as Care and Temporality. Each aspect is a figuration of the ecstatic 'unity' of futurity, having-been-ness, and the present which arises in the temporal event which founds a world.

Understanding, which is always an understanding of being, is a projection of Dasein upon its own potentiality for being, it is 'the Being of such potentiality of being'.[111] It is thus a projection of the sense or meaning of being as such, upon which are simultaneously projected beings, which are thus understood in a specific way, according to the apophantical schema of 'something as something'. In this way, beings are projected upon their being, as interpretations, ruled by the prevailing understanding of being, in its 'a priori' of fore-having, foresight and fore-conception. Such an *a priori*, in the indication of its priorities, projects horizons which enclose a *topos* of perspective for world-entry.

In this light, an *eigentlich* understanding of being would be ruled, for Heidegger, primarily by the ecstatic horizonal schema of futurity. It is an open-for as such, an opening up of horizons of possibility amid its projection, providing a conduit for this entry of beings into its world of 'binding commitments', its 'free' playspace amid a temporality of directional 'significance'. Yet, every understanding has its mood. Spontaneity is projected amid horizons of limit, states of receptivity, intimated in the reliance of the self upon an entry of beings into its world, and upon its *disposition*, this enigmatic 'schema' of mood. Each possibility, with respect to a potentiality for Being, is a thrown possibility. As this 'disclosive submission to the world', bare mood 'assails' us, it coerces, resists us as a primordial limit to one's own spontaneity of pro-jective understanding. It is in this way that understanding becomes modified with respect to the character of its own specific 'diversions', its 'interests', commitments vis-a-vis its 'current' involvements, its own temporal projections, its differing 'understandings of being' and moods.

Dasein, as its own 'there', finds itself in the mood that it has, one that discloses the attunement of the self amid its own self-affective temporality. The disposition, as a 'self-finding' (**befindlichkeit**), gives 'primal access' to the basic orientation of one's own self, prior to any expression or understanding of such 'states of being', although each mood has its understanding. In this way, mood is, in its own way, an *a priori* 'condition of possibility' for an intimate self-interpretation of Dasein. Indeed, it serves as a wedge, breach, which, amid the thrownness of existence, tears us away from absorption in the confused 'world' of an undifferentiated 'Anyone'. Mood is not only a 'way of seeing' that discloses the finitude of existence, but it is also, as a condition of receptivity, a 'way of being' indicated as *falling*, the sliding temporal movement, ruled by the 'present' and an understanding of the world which lets itself become held captive to a picture. Falling is a ceaseless failure to live up to, to fulfil, either the potentiality of that which has been, or had been promised amid a prior 'moment of vision'. Yet, for Heidegger, this is a basic aspect of our existence. In this way, radically distinct from any 'myth of a fall', *falling* is a constitutive aspect of the temporal be-ing of Dasein, indicating an oscillation within the self, betwixt its singularity and its re-absorption in the anonymous and faceless Anyone. With the foresight that that which one falls away from is its own self-encounter amid 'Nothing', falling takes myriad 'forms' of seduction, tranquilization, alienation and entanglement. It is letting oneself off the hook, a forgetfulness of the ownmost be-ing of the self, 'who' has become lost in a labyrinth of anonymity.

This indicative *logos*, playing amid the *topos* of existence, as it is constrained by these horizons of this phenomenon, is articulated as a *discourse* indicating the self-understanding of one's temporal self. We have discussed the necessity for an appropriate use of language in the attempt to disclose the truth of the being of Dasein. Heidegger seeks to lay out a *logos* which, constrained by a being, will serve to let the being be seen from itself. It will be in a turn to a radical singularity of the self that expression will have allowed for the disclosure of a being that is in itself one of silence.

The disposition of anxiety

Anxiety, in distinction from fear, is not concerned with an external or internal threat, with those modes of being which are deemed theoretical and practical, with 'objects' of the world, but with a 'threat' which is grounded in the be-ing of Dasein itself as being-in-the-world. Anxiety is a disclosure of world 'as' world. It opens up amid one's existence the inexorable possibility of impossibility.

As a disposition, fear discloses that which is 'there' for it, as an indefinite threat from outside. Fear incites alarm but exhibits the character of an indefiniteness as to its status as an 'actual' threat. Indeed it may not come, the threatening that breaks in as fear may pass away as a threat, as for instance, a disease or an imagined attacker in the night. Moreover, in addition to fear, and distinct from fear in the face of a familiar threat, there is also dread (**Grauen**), which erupts when that which threatens has the character of the unfamiliar. Yet, both of these dispositions hold in common a concern about a *merely* possible threat, one which may or may not actually occur, the threat is essentially avoidable, although the fear itself or the dread may not be so easily avoided.

With respect to anxiety, however, Dasein is anxious about its own being-in-the-world and of the obvious, gnawing threat which abides in its heart. Anxiety places the self into question with respect to its own being as mortality. One is anxious about being-in-the-world *as such*, and of the ownmost possibility of non-existence. One *gathers* an uncanny sense of 'homelessness' in this world. One does not properly belong to this world, where one has been thrown, as Caputo writes, into the 'dark night of existence'.[442] Haunted by uncanniness, one is individualized amid the vortex of its own anxiety, in which is disclosed a sense that one's 'true' world is merely 'nothing'. One flees *from itself*, from its 'world' of commitment, this free projection upon 'nothing', *towards things* in the 'world'. In the wake of this 'flight', its 'world' becomes hidden, withdrawing in the face of things. One hides away in tranquilized familiarity and in the forgetfulness only 'things' can bestow upon the one who is 'visited' by the uncanniness of anxiety.

Anxiety is not anxious about any being 'in' the 'world', any involvement, but instead, discloses this world as a *topos* of sheer possibility, as a complex projection of ecstatic temporality. In falling, one flees from itself as *this possibility*, hiding in the generic so as to avoid the intimacy of self-interpretation. In this light, 'tranquilized familarity' is a modification of uncanniness, one that 'operates' to cover up the uncanniness of 'world', of the radical temporality of 'world'. In the wake of this disposition of anxiety, the self transcends fear into a sense that its own existence, its own self, harbours irresolvable difficulties. Its own possibility of existence becomes an 'issue' for itself. In that it points to this self and not to an alibi amid the absorption of things, anxiety plays its part in this disclosure of the being of a self as Care. It is an 'existentiell' disclosure of the 'ontological structure' of one's own being. It must be emphasized that anxiety erupts as existentiell event of *a-lethea*, as a modality of the 'there' of Dasein. Indeed, anxiety is only the beginning of a journey which culminates in an explicit, though merely ambivalent, understanding of the existentiale of Care. As indication, anxiety intimates a pre-theoretical and pre-practical disposition of the self that reveals its temporal be-ing. It begins to gain a sense of its own finitude; it witnesses that 'demise' of another, and self-reflexively understands its own fate. Running away from this disclosure, it interprets this disposition, mood, according to the things of the *mere* world.

Anxiety, as a disclosure of world 'as' world, removes the veil from the average understanding of being. In anxiety, Dasein is thrown back upon itself, upon its own potentiality for being, and, in this event undergoes an ordeal of singularization. Dasein will be called back to itself, *to choose itself* – to return from its absorption in the Anyone. With the projection of the self upon its own possibility, self-understanding is transformed from that of the 'average' meaning of a 'common' understanding to one that is attuned to its own projection of world. Anxiety singularizes, tears one back from a particular falling. In a way which will be explored below, it is anxiety, and the disclosive 'movement' it engenders, that indicates 'primitive' criteria for the distinction of one's own self from that of Anyone (**Das Man**).

Care as the being of Dasein

In the closing pages of *Being and Time*, Heidegger articulates the Situation in which Dasein shatters itself against death, and amid this destruction, is forced back upon one's own singular truth, a disclosure of the temporal sense or meaning of one's be-ing. Such an attempt to find the meaning of one's existence, as we have suggested in Part 1, is fostered by the desire to uncover a 'ground', separate from that of ontic conceptualities, for a consideration of the 'meaning of Being'. In other words, in a 'retrieval' of Dilthey, and his friend Yorck, Heidegger attempts to open up a *topos* that is independent of the real and actual predication of *vorhanden* and *zuhanden*. Care may be disclosed in the context of a consideration of the totality of involvements, yet, its status as a *first*, as a ground, sets it apart from the 'truth regime' of real or actual predication. Care must have a language of its own, if, that is, it is to uncover and express the 'truth' of itself. Indeed, Care is not only 'beyond' the jurisdiction of real and ideal predications, but, as the be-ing of Dasein, originally grounds nomenclatures and language games. In this way, he contends, 'reality' and 'ideality' have 'no priority'.[443]

Care is ahead of itself in its already being amid, alongside beings in the world. Transcending, one is amid this open. Care is original 'wholeness' in which are rooted the stems of *theorein* and *praxis*, as each of these is a comportment of being-in-the-world. Even a rough sketch, however, of Care as 'being ahead of itself in its being already amid beings in the world' casts into relief, Heidegger warns us, 'difficulties' for interpreting its meaning and of its 'unity'. Heidegger approaches the existential of Care, as we have seen, in light of the distinction of the be-ing of Dasein from those of circumspective concern and of a theoretical comportment. Care, in distinction from concern, will have, moreover the character of solicitude with others. Being-with others occurs however upon a *topos* which is not oriented to the 'in-order-to', but instead to the wholeness of being in the world intimated by the *for-the-sake-of-itself.* At the same time, as the *for-the-sake-of* is intrinsically rooted in the egoicity of Dasein, there will arise a tension with respect to the who of Care with respect to the distinction between *Das Man* and the self in its *eigentlichkeit.* Care as *Das Man* aspires to a whole, yet, falls into the in-order-to in the fulfilment of its meaning, the sense of itself as a whole. It is only with the self in its *for-the-sake-of* which projects a meaning or sense of completion of Care in the disclosure of the singularity of temporal expression. The tension we are describing foreshadows the pathway from the Anyone to the *Eigentlich* self, to which we will return in the final chapter.

Heidegger underscores his search for a language outside of the frame (**Gestell**) of real and logical predication by undertaking a meditation upon the 'primitive self-interpretation of Dasein', articulated in Hyginus' 'Fable of Cura'. The Fable is a narrative and mythical projection indicating a sense of meaning with phenomenological relevance in that the latter has explicitly undertaken this difficult task to return 'to the things themselves'. The fable, as a work of art, expresses, amid its temporality of being-in-the-world, the distinctiveness of the be-ing of Dasein with respect to the myriad conflicting claims made upon its be-ing. The phenomenon of conflict in this context calls for an account of the unity of Dasein. In the myth, there is a dispute as to the name of the creature, whether it is to be called after Jupiter who gave to it its spirit or Earth who gave it its body or to Care who shaped it from the clay. It is decided by the arbiter Saturn (time) that it will exist in Care until Jupiter and Earth receive their accursed share when this mortal being ceases to live. In the 'mean time', it shall be called by its matter which is of Earth, *humus*, giving to it the name of *homo*, or man. In Heidegger's reading of the Fable, Care is the existential expression for a being-in-the-world whose being is held in question, for which this being is an issue. Care is a *topos* of possibility, a *gestalt* of human being, which is as being, phenomenon, is distinct from the comportment of indication, speech, *logos*. At the same time, it is also the 'call of conscience', a differing *logos* which breaks in, disrupts the questionable regime of predication. This is the voice of radical temporality, or, in terms of the myth, that of Saturn. Care, as a non-real/actual indication of the specificity of this be-ing of Dasein, projects a basic meaning horizon, a 'for the sake of which' upon this 'Nothing'. Indeed, Care that is evasive of its temporal meaning is a covering up of this 'Nothing' of original temporality. Amid anxiety Dasein is anxious about its ownmost possibility; its existence becomes **the** issue in terms of its 'unity', of perhaps, its *name*. It is invaded by the 'Nothing', it is 'Nothing' – shattered against death, it is thrown back upon itself, it witnesses itself as one who must decide for itself its world.

That which is vital, for Heidegger, is an uncovering of this truth of Dasein, wresting its be-ing from hiddenness and the suppression of *that* overwhelming disclosure amid the event of anxiety. Indeed, taking heed of the play of truth and untruth amid the be-ing of Dasein, Heidegger affirms that this truth, this uncovering, must occur 'again and again' amid the ecstatic play of temporal existence. It slips into forgetfulness, into the things of the world, yet, this slide itself provokes a renewed 'event' of anxiety in which the self becomes anxious about being, as it is lost in this hiddenness, in the anonymity of 'Anyone'. It oscillates amid, as Schürmann would say, a double bind. There is no escape from the truth of oneself, no escape from 'moments' of disclosure, 'events' of anxiety, which reveal the radical singularity of the self. One flees into a mere picture of oneself only to be visited again and again by the uncanny.

Since each of these temporal disclosures is, however, provisional, forgetfulness itself allows the self to be open to its changeable character that is revealed from 'moment' to 'moment'. Forgetfulness, as Nietzsche dared throughout his works, provides an opening for the creative, unhistorical 'act'. In a moment of vision, amidst anxiety, much is forgotten, displaced. This would indicate that falling is not merely a negative phenomenon, but one which provides, as the 'not', a vital and positive 'grounding' for the openness to existence. At the same time, however, there is no guarantee that this *topos* of disclosure, revealed amid anxiety, will serve as an 'opportunity' for self-knowing, much less for a deliberate attempt, as Heidegger affirms, to hold oneself in the nothing, amid the moment of vision, 'all the way to the end'.

Indeed, one may ceaselessly run away, flee from oneself, mis-interpreting these moments of disclosure as an illness, as a passing state that was 'really nothing'. One retreats to the meaning of being of the Anyone. A susceptibility to falling is indicated by its seeming inability to grasp the meaning of its being beyond these ambiguous descriptions, where even phenomenological indications such as Dasein being its own 'there' becomes reified as 'theory', as a 'play amongst mere concepts'. We no longer look towards the phenomenon itself as we have become ensnared in the detour of discursivity. Care is however already ahead of itself, and always coming back towards itself as it fulfils its projected 'potentiality for being', but, like a construction site, it is never complete, 'At each moment', Bataille suggests, there is the 'impossibility of a final state'. The meaning of Care, however, is beyond itself in temporality, which in Bataille's terms, is disclosed in the step back from domesticity to the wildness of *ipse*.[444]

The incompleteness of Care

The ambiguity and falling which is a potentiality of Care is 'grounded' in the dissemination of a disposition of anxiety amid the exigencies of everyday life. The being of Care is a for-the-sake-of-itself ceaselessly on the outside, transcending, already ahead of itself and alongside beings in the world. Its 'meaning' is temporality, and, as I have suggested, there is a marked tendency for Dasein to flee itself, to suppress this anxiety, hide away amid this anonymity of things, 'within the world'. This hiding provokes more anxiety, however, and one awakens to understand that from this anxiety, and from that which it discloses, there is 'no exit'. From the perspective of its flight to entities in its world, Dasein understands itself according to these entities, and thus, its own meaning is gathered from an understanding of being of circumspective concern, of that which is ready to hand. In this light, the be-ing of Dasein as 'Care' would have no meaning beyond that of ceaseless but impossible striving after a not-yet, life as an utter absence of completeness, each never living to see 'death', but, hurrying to produce eternal monuments. A corpse is left there for the 'other', cremated or buried. Heidegger contends that such a state of ontic incompletion and those various ontical reactions against it, do not give the true sense of 'unity' for one's existence. Indeed, it is a fragment of an ontic domain of beings, passing over the intimacy of the sense or 'unity' of being which is intuited in the radical singularity of the self.

The notion of a general purposiveness, in other words, which is laid out in Care is not sufficient to account for the meaning of this phenomenon. It is 'incomplete' since it is placed in isolation from the context of its origination in anticipatory resoluteness. Isolated, it becomes assimilated to an understanding of being of that which lies closest, ready to hand, as practical, as purposiveness without purpose. When a breach occurs, in this state of incompleteness, there will emerge theoretical attempts to make sense of this undifferentiated striving without any clear direction of its own. With such a will to a system, one that was declared impossible by Gödel in his own way, it would not be difficult to begin to understand notions such as the 'determination of the will' by reason and its regime of pure practical precepts of reason, or the subsumption of imagination under the categories of pure understanding, in the manner of Kant. In itself, Care, as the be-ing of Dasein, is meaningless, as being-in-the-world only, it is non-sense. Without a sense of being, the for-the-sake-of one's own singular self, it will be given a tacit ontological commitment in its absorption to the regime of predication. This ensnarement into Reality forecloses on the attempt to grasp the meaning and truth of Care. Truth, in this context, indicates an 'event' in which there is a sense of one's being, one which may grow deeper, if one refuses to flee this intimate self-disclosure amid these horizons of finitude, of temporality. One falls from this specificity of Care, one forgets the 'ontological difference' that distinguishes Dasein from that of entities 'in the world'. Dasein is 'in the truth', it is its own 'there' (**Da**). For Heidegger, there can be 'truth only in so far as dasein is and as long as dasein is'.[445] Primordial truth, distinct from 'truth' conceived as mere 'logic', is disclosed in a retrieval of a meaning for one'self as an abyss (**Abgrund**), one that opens up as lived anxiety amid temporal being-in-the-world.

In the final chapter, I will set forth this limit Situation, this 'glance of the eye', or 'moment of vision' (**Augenblick**) of anticipatory resoluteness. Indeed, anxiety is only the beginning of this disclosure of an understanding of one's own be-ing as Care, and the attempt to articulate this singular sense of one's being. One is called by oneself back from its absorption in the 'Anyone' to a temporal meaning of oneself, the singular 'truth' of oneself witnessed in this moment of vision. One is Guilty! in Anaximander's sense, as one is inescapably temporal, just as are the others with whom I dwell. It is one's acceptance of this guilt, and the 'silent' understanding of one's potentiality for being amid this predicament that gives one an understanding of the meaning, or unity, of one's own be-ing.

Chapter 13

Temporality as the Ontological Meaning of Care

But because philosophy is the most radically free endeavor of the finitude of man, it is in its essence more finite than any other.[146]

In the opening pages of *Being and Time*, Heidegger intimated the possibility of temporality as the 'transcendental horizon for the question of the meaning of being'. He confirms this intimation in Division Two, 'Dasein and Temporality', where he indicates temporality as the meaning of the being of Care, one that is disclosed through an 'existentiell attestation' of an 'authentic potentiality for being a whole'. The pathway for this uncovering is marked with the existential characters of 'being-toward-death', 'call of conscience' and 'guilt', whence each indication has been expressed to point out that which is disclosed amid an event of original, ecstatic temporality. To think along with Schürmann's reading of Heidegger's *Contributions to Philosophy*, for a moment, each of these phrases expresses an aspect upon an existential *topos* which shelters an event of singularization.[147] Each of these words is a mark which traces the being of this event of temporalization, of the *to-come* of Da-sein as that which emerges amidst the discordant eruption of a 'world'. At the same time, however, this event is not simple, nor need it occur in an instant, conceived as a 'now', but can be of a quite longstanding duration and intensity. It is the seizure of the self by a *topos* of de-cision, in which one must decide whether or not to heed the call of singularity, whether or not to choose the *eigentlich* self. In this way, Da-sein is *not-yet*, but may come to be. Heidegger has sought to disclose his makeshift thinking, one which is suspended over an abyss of no-thingness, and has laid out a possible pathway for the disclosure of the meaning of singular existence. In light of fallenness of the Anyone, however, it is just as possible that he will be misunderstood, and, with Cassandra, will stand silent, forced to witness the tragedy *to-come* before his eyes. His words will perhaps be given the same regard as those of Zarathustra in the marketplace. Yet, there is also always the *other* possibility.

There has been a good deal of misunderstanding surrounding Heidegger's use of these particular words, death, conscience and guilt, in light of their usual religious connotations. Those familiar with his method of departing from everydayness with an access-definition of pre-philosophical understanding, will understand, however, that the indications of death, conscience and guilt are a recollection from historicity which will seek to retrieve the *eigentlich* meanings of these terms. He resists theological, ecclesiastical resonances to his indications, expressions. He rules out from the start any reference to a 'morality of obligation', of an origin in calculated debt, in the regime which Nietzsche sketches in his *Genealogy of Morals*. Such a reference would be to fall back again into the theticism of the 'Anyone', one guided by the 'logic' of entities. In this way, for Heidegger, the 'official' translation is at best ambiguous. In other words, the *way* Heidegger's philosophy is *translated*, interpreted and expressed, will not only decide the 'metaphysical' commitments, haunting certain words, but will also be susceptible to appropriation by the theoretical logicism and theticism which holds sway 'in the present'.

The self exists within a horizon of temporality, amidst the double bind of natality and mortality. The specific character of mortal existence, with respect to the traditional interpretation of being, is that of a fugitive being within the world, where death is seen as a debt long overdue. This interpretation of mortality as debt intimates Heidegger's criticism of the faulty translation of Anaximander by the tradition. The exposure of these faults will aid us in our understanding of Heidegger's attempt to retrieve a more authentic meaning for words such as conscience, guilt and resoluteness. He criticizes the translations, for instance, of 'coming to be' and 'passing away', 'pay penalty' as appropriate interpretations of the expressions *genesis* (γενεσις), *phthora* (φθορά) and *dikia* (δικία) , with respect to their own indigenous expressive historicity. With some irony, Heidegger states that he agrees that the 'usual' doxagraphic translation is 'literal'.[448] Yet, he attempts to evoke un-tapped possibilities, covered over, suppressed by a historical 'regime' of interpretedness, one with its own tacit and explicit philosophical commitments. In other words, this usual translation neither gives Anaximander the chance to speak for himself from 'out' of his own historicity, nor allows this event of be-ing to speak through him.

In light of the case of Anaximander, Heidegger must begin in the everyday as a *point of departure* for our pre-understanding, yet, with respect to the *beginning* in the projection of world, these historical terms must be deconstructed, excavated, so as to prepare for a retrieval which seeks to understand the meaning of this λογος, to see that to *which* it points. In this way, the scenario of debt and guilt which is intimated, yet, denied, as a possibility in each of these artefacts, serves as an exemplar of what Heidegger specifies as an 'inauthentic' understanding of the being of Dasein as Care. One departs in the un-ownly, thrown into existence in the middle of a sentence. We proceed from the vernacular of death, conscience and guilt, and guided not only by Heidegger's rejection of the morality of debt in *Being and Time*, but also by his other readings, travel along the path that leads to his interpretation and expression of these 'events' from the perspective of the radical singularization of this self. Such an indigenous self-interpretation will not only disclose the singularity of one's self amid a 'moment of vision', in which one is called back to itself from its lostness in the 'Anyone', back to one's limits, in a 'glance of an eye', but will also lead to an acceptance of this finitude, which is, at the same time, a coming to a stand for oneself, where one's 'potentiality for being' becomes an 'issue' for itself. One builds one's makeshift thinking, one's world, homeless but sheltered, a stranger thrown from 'nowhere'. Circling in a thinking which finds its point of *departure* in the vernacular, it comes to find its own *beginning* in an intimate self-interpretation and free expression of oneself, in a retrieval of meaning for one's ownmost being, expressed in a language, *logos*, appropriate to this free being-in-the-world.

As with the narrative of the returning philosopher in Plato's 'Allegory of the Cave', one is told, amongst the shadows of the cave, of a realm of light, of illumination, one is given a map to find one's way, a sketch with the markings of significant junctures. Yet, unless one has had the temporality of one's being disclosed in a moment of existential vision, of the 'event', as Wittgenstein[149] and Nietzsche[450] intimated, such talk of an opening of light would surely go unheeded. Heidegger himself is attempting to give to his reader a 'rough sketch' of his path of being-there. Yet, for those captivated by the absolute 'Reality' of this dimension of shadows, such talk of a beyond, of a realm of light, is seditious, resembling the lies of the poets who have been cast out, banished beyond the city walls. These philosophers are silenced, or, at least the attempt is made to do so.

In the following pages, therefore, we will trace this pathway of the radical singularization of the self, an event of the self amid its limit situation and 'passage'. Yet, this is an event, retrospectively, which must find itself articulated in a discourse that divides 'things' up into 'momentary sites' or places along a pathway, from an existentiell disposition of anxiety in one's being-towards-death, to a call of conscience, and of one's being guilty, rooted in a moment of anticipatory resoluteness. Although such an event may be an existentiell disclosure which seems to erupt amid an 'instant', Heidegger retrieves this temporality of anticipatory resoluteness *phenomenologically*, describing this event as a 'spiral', a descent into one's own abyss, into *my* maelstrom. Yet, this is a spiral where at each 'turn' of possible self-disclosure, there obtrudes that seduction of escape, of an evasion of the singular meaning of oneself, of the truth of oneself. One is tempted to recoil in the face of the nothing, of this existence of being thrown 'in-between'.

In a descent without evasion, however, one's 'binding commitments' are projected as a horizon which is *a priori* with respect to any beings which enter into its 'world'. Yet, this *prior* is itself temporal in that the projection, emerging from facticity, is itself projected upon temporality, and thus, one's world, projected as 'binding commitments' is susceptible to transfiguration. From out of the facticity of beings, the self *becomes* – it is not-yet, it is never-yet. The self, in its thrown understanding of being, projects the horizon for a topology of beings.

The temporalization of the facticity of resolution discloses the possible transition to a metontology from out of the heart of fundamental ontology itself.

The extremities of the self

The being of Da-sein (being-there) has been indicated as Care. It exhibits the organized '*topos*' of the phenomenon. As I have suggested, however, Care, inspected in isolation from its temporal ground, cannot give an explanation for its own 'unity'. That which is lacking in the indication of Care as the *there* of thrown projection is precisely the freedom of the abyss, of a makeshift 'ground' that necessitates that one freely appropriate one's binding commitments, one's world. Care, as the *topos* of thrown projection is already ahead of itself amid its dispersion into entities. With respect to this question of freedom, such a *topos*, where a self merely cares, descends along a spiral of ultimately unfulfilled projections, gives, as I have suggested, no explanation for the 'unity' of Care. Neither the relevant phenomena, nor the *logos* or expressive orientation come from the ready-to-hand (**zuhanden**) or present-at-hand (**vorhanden**). This question is that of an ownmost or singular wholeness or 'unity' of Da-sein, of the meaning of this be-ing of Da-sein, and a provisional 'answer' expressed in a mode of discourse that is 'attuned' with the being of Dasein as Care. Heidegger's own 'answer' is, as we have seen, the Nothing (**Das Nichts**) which one encounters in the mood of anxiety. This would seem in the first instance to be incompatible with the aspiration towards a 'wholeness'. Indeed, anxiety sets us off into flight, to hide 'in' beings, a concealment which, in its turn, instigates a spiralling amplification of anxiety, a vortex of the questioning self. How is such a situation compatible with a 'possibility of a whole'?

Anxiety, as a disposition which discloses this be-ing of Dasein, must have dwelling within it, at least implicitly, an 'understanding' of the circumstance of be-ing that is being disclosed. That understanding is indicated in the existential character of 'being-towards-death', one of the most prominent schemata of falling, but also, of the end, of the limit of existence. Death is the end of being-in-the-world. Death is the limit horizon for any possible 'meaning' of existence. This understanding as a projection is an indication of a phenomenon of its own temporal directionality, not in terms of some thing that is feared, nor that which moves down a time-line, but a 'necessity' that is always 'there', as an intimate possibility of a self, as it carries its death, its 'end', along with it. But implicitly amid one's own temporal existence, there is indicated a 'task' by which one brings to clarity, to explicitness, one's own 'understanding of being'. This task is enacted as a retrieval of the meaning of one's own being through the 'basic' disclosures of 'truth' in the event. Such a 'task' guides us along a path through which one moves into clarity. 'It' occurs 'at once', yet, each disclosure is a touchstone along the way, reminding one of 'what' one has been, although any 'moment of realization' could be forgotten, or, become distorted in the 'primo-rdial associations' of our 'surrealist' temporality.

The question of the meaning of Care, of its being-a-whole, and thus, of death and temporality, if conceived ontically, leads us into paradox. Care is presup-posed to have an end but an end that is its own dissolution. Indeed, Care can never be whole, when its 'fulfillment', at its 'end', is that which 'destroys' its own being-in-the-world. The paradox is that as this self reaches its 'end' it simul-taneously ceases 'to be'. In this light, its own most 'meaning', its 'being-a-whole' perhaps is 'really' a 'nothing', merely the demise of an 'animal'. Its 'meaning' is a corpse, but only for the other. It crosses from this being of Care to the being of the present-at-hand, without any significance, a 'tale told by an idiot signify-ing nothing'. From this perspective, the 'end' that is 'death' is my being-un-alive. Yet, I will never 'experience' my own death as a corpse 'there' – it is only the death of others that discloses an ontic, existentiell 'example' of 'death'. Heidegger insists however that this question of meaning be asked ontologically and that the be-ing of Dasein in regards to the task of articulating an 'end' that is appropriate to its be-ing be considered in *its own terms*. Such an 'end' would neither be that of a theoretical 'totality', 'systems' and 'processes', nor of practi-cal beings in eschatological projections of meaning. The meaning of Care must be disclosed in an intimate self-interpretation of lived, temporal existence.

Heidegger has entered into a phenomenologico-existential exposition of death as a search for an 'authentic being a whole' of Dasein, of *my* self amid its world. From the 'outside', *that* 'authentic being towards death' of this one who dies cannot be 'experienced' by myself. I stand before this still one, beside her fresh grave, but can only sense *my own* being *toward* death. Although I remain haunted by the image of her grave, and of her face, I cannot step beyond this horizon of non-substitutability, of non-related-ness. Each is singular with respect to an intimacy with death, no one can experience the 'pathos' of that other amid her 'loss of being'.[451] Heidegger announces:

We are asking about the ontological meaning of the dying of the person who dies, as a possibility of being which belongs to his being.[452]

We look at this corpse, we imagine that we are 'there'. We stand in the place of the other, this one who has 'met his demise'. Yet, in this impossible exercise of sympathetic representation, we succumb to the limits of substitutability, or translatability, as we cannot 'know' the death of another: 'No one can take the Other's dying away from him.'[453]

Heidegger contends, in the darkness that surrounds the other as an *aura*, that one can understand one's 'authentic-being-a-whole' only from amid *one's own being-toward*-death. The paradox is evaded in the ontological-phenomen-ological indication of the ecstatic character of temporal existence, and of the projection of the ownmost being of the self upon the ecstasy of the future. It is

not demise, but a horizon of possibility that is disclosed. In light of this phenomenological analysis of death, the sense, or meaning, of a 'being-a-whole'
shifts, which, contrary to this silence of a 'corpse', is disclosed amidst the event.
In this way, death is to be understood as *my* being-*toward*-death, as a projection
of futurity from amid the *topos* of my being-with-life, in-between. Amid being-
towards-death, which as a situation of limitation, Care initially is uncovered as
the be-ing of Dasein. Care as existence grasps its own possibility of being as an
'issue'. Amidst its being-towards-death it encounters its 'ownmost non-relational
possibility',[454] *perhaps*, it comes face-to-face with itself, with its mortality, one 'not
to be outstripped'.

Yet, amid everydayness, this self covers up its face-to-face with itself, it flees
in the face of itself, its own finite be-ing, falling away from its ownmost 'truth'
into that of practical *mores*. Indeed, the self which comports itself to itself in
everydayness is the self of the 'Anyone'. In the everyday, public death is inconspicuous, as if death were not yet present at hand (**vorhanden**), for a 'nobody',
always someone else, not me, not yet . . . levelled into evasive concealments,
tranquilized familiarity, death 'disappears' or becomes the solemn spectacle.
Anonymous strategies of the 'Anyone' orchestrate our comportments to death.
They work to turn our anxiety into fear, only then to dismiss this as weakness.
Not to 'permit us the courage for anxiety in the face of death',[455] we are
'embalmed' by the 'funeral home industry'. In this removal, an 'absense' of
death, the illusions of a superior indifference seduce us towards an alienation
from our own existence, as each of us, despite our 'a priority', persists amidst
this play of a basic 'untruth' or concealment, our temporal being of falling.
Yet, even in this flight, in this covering over, the denial of our being-towards-
death, there is a tacit reference to this 'phenomenon' of being-towards-death.
'Anyone' suppresses, covers over this 'at the moment of death' and deflects,
suppresses this anxiety into our fear amid this world of concern, as a 'will to
survival'. In its being-covered-over, death, an uncanny guest, loses its 'certainty',
inevitability. One loses this tragic chance, possibility, of an indigenous expression of its singular, temporal being-in-the-world.

In being-towards-death, Dasein has, writes Heidegger, already decided itself
'in one way or another'.[456] Amid a disposition of anxiety via which one's finitude
is disclosed, it is possible that one may turn away from one's ownmost possibility, fall into evasion, cover up this being, suppress this anxiety, lose oneself in
the plethora of things within the world. Anxiety (as with death) is managed,
organized, produced by professionals; it becomes calculable, profitable. For the
ones who orchestrate death and anxiety, each self is expected as something
present at hand. Heidegger contends that this weakening of the awesome
character of death via 'calculation', 'commercialism' and concealment is a suppression of the openness to an encounter with the mortal singularity of one's
self. Such understanding concerns not any 'actuality', but this closeness, where
one 'fathoms' that possibility for an impossibility of existence. Death gives

us nothing. Resisting the evasive temptation to reify our being-towards-death, such being becomes 'greater and greater as the possibility of impossibility' is uncovered as a 'measureless', in a similar sense as the Unlimited of Anaximander. Being-towards-death discloses one's singular potentiality for being in the self-disclosure of an anticipation of impossibility. Indeed, one's be-ing is disclosed as anticipation itself, as the *predicament* of one's own existence. This self-disclosure 'wrenches' one away from the dizzy-ing proliferation of entities, turning back to one's finite self in its freedom, toward one's singular possibility and meaning. This is not a narcissistic retreat 'into the self', as death 'lays claim to dasein as an individualized dasein'.[457]

This 'there' of one's existence is disclosed through this event of singularization. This singularity is the honesty of this self before itself. Heidegger intimates that in this 'moment of vision', although one never leaves one's predicament, practical concern and solicitude fall away as this self projects itself upon the 'possibility of its impossibility', it is singularized from the 'Anyone', the self stands naked in front of itself. Amid the event, one is 'coerced by the phenomena', forced to take over one's being, to accept this possibility of its being (although one can cover it up later). Free for one's own death, Dasein is liberated from evasion and illusions, awakened to the facticity of its freedom to choose its possibility, one which lies 'stretched out' as one's ownmost possibility, and in this anticipation, one is also free to transfigure one's being, to 'shatter all tenaciousness to whatever existence one has reached'.[458] This free-projection is a temporalization 'out of the future', amidst which one finds a way to take existence as a whole, in which being-a-whole has significance vis-a-vis this distinct be-ing and expression of existence. The singular self, Heidegger writes, anticipates all possibility, but for only an 'instant' and certainly not as anything 'actual'. One comes to oneself via projective acts. Betwixt the plays of temporality, these moments emerge and dissipate, they rise and fall. Each 'moment' is a piece-work disclosure of 'historicity of the self', a nexus of temporally self-deconstructive 'binding commitments', a provisional expression of truth and meaning.

The disposition of anxiety discloses the 'constant threat' of one's own 'there'. *I* am brought 'face to face' with *my* nothing in *my* being-towards-death. Amid this possibility of non-relation, non-existence, it would be appropriate that this ecstatic self seek its meaning of being from amid its own free-projection, from its being-in and its self-expression. 'Common sense', 'logic', *break down*, as there is no 'sufficient reason' in being-towards-death. This awakening of the possibility of impossibility intimates our previous exploration of imagination as that which conjures a presence of absence, and as the seat of these temporal projections of finite knowing and existence. Heidegger contends that we must abandon the 'reason' of the tradition, and must be seized by original temporality, which is, as he asserts, the truth of a reason that, as a faculty of principles, is at the same time, a faculty of imaginative synthesis in the 'schematism of the concepts of understanding'.

The indication of singularity is not meant, however, to suggest a scenario in which a subject keeps itself away from the 'world'. Instead, Heidegger discloses that being-oneself 'takes the definite form of an existentiell modification of the "they"'.[459] One begins in everydayness, in the Anyone, one makes no choices, at first, carried along in the 'nobody' and 'everybody'. Yet, the fall of the self into its obscurity in this tranquillity of familiarity, being a possibility of the temporality of existence, will transfigure itself through its reversal into a bringing oneself back from a lostness in anonymity, evasion, to this 'choosing to choose', a finding of oneself in the disclosure of one's being-there. It does not 'have' itself, it is 'on the way' to itself. The reversal is instigated by the uncanny call of the self from nowhere, from its own self as a possibility. In the moment of anxiety, in which the truth of the self is disclosed, previous worlds, commitments, just as the things which entered into these worlds, fall away as one holds oneself in the nothing. From out of this encounter with the nothing and the suspension of one's world, another possible world may emerge, the possibility of which is expressed as a 'call of conscience' (**Gewissen**). Again, it is not Heidegger's intention to intimate a 'theological' significance to his indication. Indeed, conscience can be interpreted literally, as a finite self-knowing. In this shift of aspects, it is from the interpretedness of the Anyone, of this thrown space of departure, that one is called back towards oneself, called to know oneself via a disclosure of one's limits. Amid anxiety, in this moment of vision of being-towards-death, in which one is cast away from 'normality', the self may immediately suppress such a torment by means of evasion, deflection, sublimation, narcotization. Yet, to this 'call of conscience' corresponds a 'hearing'. This voice of conscience 'gives us something to understand'.[460] Mood unearths the being of the self; anxiety discloses this being as Care, which understands be-ing as *my* being-towards-death. This voice beckons me to fathom, through my being-towards-death, a 'world', a meaning for this be-ing whose fate is death. One seeks to evade this timebomb of one's being-towards-death, one runs away from the vertigo of this 'perverse' contemplation, and flees into the 'world'. Yet, there is no escape from this 'spiral' of anxiety as one slides, oscillates between nothingness and the 'Anyone'. One hears, listens to oneself, lets another hearing break in, one's 'binding commitments' undergo 'existentiell modifications'. The 'call' is an 'abrupt arousal' of a 'giving to understand'. Heidegger poeticizes

The call is from afar unto afar. It reaches him who wants to be brought back.[461]

That call initially reaches the 'Anyone' but as it is addressing a singular self, in that the self is being called back to itself, the 'They' collapses 'in the appeal'. One is called back to oneself, 'where' there can be no generic 'one size fits all' theories and normative prescriptions. In the call, these latter 'constructions' fall into insignificance, veils fall away, the self stands amid this intimacy of its own existence. Being-open to the raw 'matters themselves', the call 'discourses solely and constantly in the mode of keeping silent'.[462] As I have mentioned, the openness, the truth, begins to disperse with utterance, it becomes lost in the 'detour' of discursivity. This 'call' throws the self into a disposition of 'reticence over itself'. This self as da-sein 'calls itself'.[463] The self calls itself to witness this be-ing of its ownmost possibility. This is, however, not 'something that is planned', as if it were a meditative exercise initiated by a self to occur at a pre-arranged time. Indeed, Heidegger writes,

'It' calls, against our expectations, and even against our will . . . [464]

The call, in that this self is thrown, comes 'from me and yet from beyond me and over me'.[465] In its silence, da-sein is disclosed for itself simply in 'that' it is, not yet with respect to 'why' it is. In this light, conscience, as a disposition of the being of a self in its thrownness, discloses a self in its finite existence. With anxiety, in a similar way, Dasein is brought before the nothing into the depths of uncanniness. A self that is singularized, down to its own 'nothing', discourses to itself in a voice which is alien to the generic 'Anyone'. For all practical and theoretical purposes, it is silent.

In a call of conscience, the discourse of the self is silent, it releases itself from reportage, allowing itself to disclose, express itself, in its potentiality-for-being, to call 'itself' back from its falling. Heidegger lays out a scenario in which one is pursued by one's own uncanniness, writing '"it" pursues me, I pursue myself'.[466] As with anxiety, the call of conscience is a disclosure of receptivity, disposition, which 'gives us something to understand' with respect to this be-ing of *my* self. That which it gives me to understand is that my existence is that of being-guilty. Once again, guilty (**schuldig**), is not meant in a 'theological' sense, or in the sense of a 'morality of obligation', but instead, with respect to whence and what a call 'gives to understand'. 'I' am called back to myself, to my 'there', to remember myself amidst the chasm of thrown singularization. Heidegger insists that while we can begin with the everyday understanding of guilt as debt, especially in regard to the everyday meaning of substance (ουσια) as holdings or assets in an ontical metaphysics, 'the guilt of concern', that we must instead look elsewhere for the *ontological* meaning of being-guilty. Being is, as Nietzsche intimates, an 'innocence of becoming'.

The everyday understanding of debt is that of a lack of payment with respect to another. In a *destruktion* and appropriation of this pre-understanding, Heidegger articulates a formal conception of being-guilty as understanding oneself to be the basis of a lack, and moreover, understanding oneself to be a 'lack as such'. 'Ordinary' phenomena, such as debt, 'drop out' when Dasein understands itself as 'nothing', as the 'not'. Being-guilty, in this light, points to an openness to oneself, to myself 'being the basis' for oneself, to accept this self-grounding amidst one's thrown existence. Dasein, this self, is its own thrown 'ground' for its potentiality-for-being. It projects itself onto possibilities, given amid the horizons of thrownness. It can never overcome this thrownness, it must *be* its own 'there'. This self, however, thrown to itself, already lags behind its possibilities. The self is not therefore its own substratum or some theoretical ground, but is being-the-basis of itself, as its existence, of its own nullity. This self is not a thing, but a projection upon a potentiality-for-being, 'outside' of itself amid ecstatic temporality. The appeal, the call, is 'issued'. And there is a hearing, Heidegger writes, that is attuned to the appeal, if it is to be heard from itself. In this debt, one is guilty in that he or she is mortal. This usage, as with Anaximander's fragment, differs from an ordinary usage of guilt, in which there is a transgression, followed by a determination of guilt. Heidegger insists however that existential guilt is primordial. He suggests that we can be open for our guilt and that we must want to 'become guilty', we must accept this guilt. At the same time, by choosing these words, Heidegger is re-casting these historical biblical words as fallen expressions, ones which need to be retrieved amid their original horizons, or, in other words, to locate this temporal 'ground' and 'falling' of these conceptions of guilt and conscience amidst the *eigentlich* coming to stand of Heidegger himself out from his own nothing. In this light, guilt is being-temporal, wanting to be guilty and wanting to have a conscience, it is my own decision to be open to, to accept, my temporal be-ing, to accept my limits, to 'know myself' from amidst the 'limit situations' which seize hold of my existence. The everyday understanding of guilt as a morality of debt suggests Kant's critical regulation of the will via reason in its practical deployment, pictured as a court of justice. Heidegger contends that such a conception of conscience misses the 'innocent' content of the intimate will as the eruption of an indigenous self-understanding and expression, as the call of Care amid uncanniness, an existence that returns to itself, bringing 'reck' into its being.

In choosing myself in this situation as this self disclosed in this anxiety of being-towards-death, 'I', after Heidegger's manner, have projected my world from out of nothing with resoluteness, choosing an authentic existentiell potentiality for being-in-the-world. Resoluteness is a retrieval of the openness of Care, and the disclosure of the *for-the-sake-of-which* of the *eigentlich* self, away from the interpretedness of the Anyone. A retrieval of the self in the sense of a projection of oneself upon one's own being-guilty implies a re-orientation of one's received commitments in that Dasein has freed itself for its 'world'. It cannot overcome thrownness, but it can awaken a possibility of itself and decide for this possibility. That which is specifically resolved by a self is disclosed in a resolution which delimits the situation. Dasein, the being of the there, the situation is 'founded upon a resolution',[467] and it occurs amongst 'Anyone', as a modification of the 'Anyone'. The limit situation is the 'there'; it is disclosed in resoluteness. In this way, resoluteness is a temporal *a priori*, projected as a horizon for an orientation in one's 'world'.

Resoluteness, as a disclosure of and call towards the being-guilty of Dasein, to this being-temporal of a self, uncovers its potentiality for being of its self amid the anticipation of its death. It is amid this event that the self not only witnesses its being-thrown into its ownmost possibility, but also, as free, is called upon to decide from among this plethora of possibilities – **which** possibility. It is in this way that the binding commitments of its world become 'established' in their original free projections. Such an original temporality, as a temporal *a priori*, or, instead, with Schürmann, an *originary double-bind*, for existence, discloses this self as 'at bottom' as a *topos* from which there is a dissemination of differing 'modes of the temporalizing of temporality'.[468] As with anxiety and conscience, anticipatory resoluteness discloses the truth of one's temporal self to itself, not only as disclosure of the being that it is or has been, but as that which can be, amid this 'glance of the eye', of de-cision, of freedom without evasion. Yet, this is truth which is to be understood in relation to original temporality, and as it is 'truly' open to existence, we must be prepared to 'take back' our resolutions and keep ourselves open for change. Heidegger writes:

> On the contrary, this holding for true as a resolute holding oneself free for taking back, is authentic resoluteness which resolves to keep repeating itself.[469]

The 'resolute holding oneself free' is an openness to temporality and to the makeshift character of one's thought and being. One's thoughts and practices are intrinsically temporal and able to be given up in that our thinking practice is guided by the 'matters themselves', which as temporal, are ceaselessly calling forth new self-expressions of existence. This freedom, in its attunement with the phenomenon, discloses a self in its truth, in its radical temporal existence, displacing the denial and the *cover ups* of the 'terrible truth'. Heidegger warns us, however, that such freedom, in that it exhibits a destructive orientation, 'has the character of doing violence'.[470] He asks, however, how are we to contrast the violences of his own 'counter-ruination' with the initial 'ruination' (and incessant repetition) of 'tranquillized obviousness'. For Heidegger, violence must be acknowledged in its 'event', just as most approve, without hesitation, the 'necessity' of the 'violence' of Descartes or Kant, or, of Socrates, as he corrupted the youth of Athens and openly worshipped foreign gods. If Heidegger is committing a 'violence', then he could argue that it is a violence 'coerced by the phenomena', by a 'call' to throw off falsifications and evasions, to embrace one's finite possibility and encourage others to pursue openness. Common sense, he says, must pass off as 'violent anything that lies beyond the reach of its understanding or any attempt to go out so far.'[471] He would not 'equate' however violence against phenomena with his violence against constructions of 'theory', which as mere pictures of the world, are violent in themselves as sheer ruination (**ruinanz**).

With anxiety, as we have seen, Care is disclosed as the be-ing of Dasein. Care is a 'structural whole', a 'totality' for the being of the one whose existence is an 'issue' for itself. The 'unity' of this being of Care is not that of a 'logical principle' nor a conceptual 'unity' of subsumptive judgement, but is , as we have seen, a unity in the sense of an 'ontological unity' amid its many comportments or ways of being, a 'unity of being', an 'orientation' for this holistic lived *topos* of Care. Further as the call calls the self to be open to its own temporal being, the question of the meaning of Care is at the same time the question of the 'unity' or meaning of this singular self, as Care 'harbors' this self.

Contrary to the bare 'I think' which Kant posits to 'accompany' each of his representations, Heidegger insists that the self can be understood only with respect to its being-in-the-world, which in the language of Kant would be its 'phenomenal content', or the makeshifts that do not matter in Husserl. Heidegger writes, 'In saying "I", Dasein expresses itself as being-in-the-world'.[472] With the 'I' of the Anyone, the phenomenal self is overlooked in that the non-specific character of theoretical categorization manifests itself as inappropriate for the formal indication of the phenomenon, or for allowing the phenomenon to show itself. In this way, the 'I' can be made to point to something else besides a logical, numerical 'subject'. With the 'I', 'Care expresses itself', amid the naked singularization amid its anxiety. The 'unity' of this being of da-sein, of the self in its being ahead of itself, already alongside beings within the world, is, as Heidegger insists, meaning which is 'there' only amid an anticipatory resoluteness, in one's openness to one's temporal existence. In this way, the 'unity' of the self, as the 'constancy' of the self, is 'grounded' upon an acceptance of this temporal meaning for one's being, but also upon these myriad projections, as each transcends towards one's possible futures. Yet, this 'general' schema of the meaning of 'unity' risks the reification of theoretical generalization.

Indeed, for Heidegger, the 'ground' of 'unity' or meaning for Care lies in this resoluteness itself, in what one witnesses in one's anticipation of death and of the trauma which occurs in the attempt to fathom one's own meaning, or to cover over this 'breach' erupting in the event of disclosure. In this possibility, one fathoms the radical limits with respect to that meaning of another Dasein, another self, whose secret is vis-a-vis someone else a 'placeless center of presencing' of this *a priori* world (κοσμος), into which enters its 'world'. As with anxiety and conscience, anticipatory resoluteness as an event, amid the disclosedness of the 'there', discloses a self, and, one not only fathoms the temporal meaning of one's being, but, more specifically, one resolves, amid this situation of freedom, to be faithful, open, to this meaning of one's existence. Upon coming back from this 'event', 'into the world', one carries this self-given resolution along 'with' oneself. It 'abides' in one's lived existence, as its makeshift 'a priori'. In this existentiell attestation of an authentic potentiality for being, temporality is disclosed as the 'ontological meaning of Care'. Heidegger writes:

Meaning signifies the 'upon which' of a primary projection in terms of which something can be conceived in its possibility as that which it is.[473]

Amidst the event, the self projects its existence upon the ecstasis of the future, upon its potentiality for being, thus uncovering temporality as the ontological meaning of Care, as the meaning for this being of the self. Care harbours a self 'as long as it lives', as it comes towards itself (phenomenology of aging) from its finite ecstatic projections. The ontological meaning of Care, or temporality, is the self-understanding of a self, as it comes towards itself from the future. Yet, he 'moment of vision' of a 'future which makes present in the process of having-been' is this singular event of the situation of disclosure, in which the 'unity' of one's being is disclosed as a temporality event. This 'unity' of the ecstasies is made explicit in the 'limit situation', as the 'unity' of self-projection upon a 'for-the-sake-of-oneself'.

The ecstasy of silence

The specific temporality of this 'moment of vision' has already come into the foreground in light of the question of the temporal status of these existential characters of *Being and Time*. The occasion for this question is given by Heidegger himself who referred to, in *The Metaphysical Foundations of Logic* to the 'extreme model' of *Being and Time*, section 64, noting the 'unphenomenological' rigidity of the self-projection of Dasein in this resolution of its 'moment of vision'. It is in this way that we must explore the context of his project of radical phenomenology, espoused in his lectures on temporality. Otherwise there is the risk that one might overlook the radically temporal character of Heidegger's early philosophy. Levinas[174] seems to have mis-read Heidegger in this way, castigating him to move 'beyond essence', even though it was the latter who had already issued a call to move beyond both essence and existence in his indication of temporality as the transcendental horizon for question of being. Heidegger is depicted as an 'ontologist' who could not get beyond a 'doctrine of essence'. It is this type of portrayal which would lead Heidegger to drop the expression 'fundamental ontology'. Yet, as he makes very clear, *these existentiales are his own*, they are what he has taken back with him from his own 'moments of vision'. This is his own temporal expression, a 'makeshift' that points to, and expresses his 'world', the phenomenon of his own existence, 'begun each time anew' amidst his temporalizing of temporality, which, as Dasein, is a being that in each case is **mine**.

Radical phenomenology can only be honest, as I suggested at the beginning of this study, if it is 'grounded' upon an ontic fundament. Heidegger points to the phenomenon of his own finite self upon its precarious 'ground', in a thrown temporal self-projection amid being-in-the-world. He seeks to uncover the temporal abyss of the philosophizing self in his or her temporality, whose sense of being is uncovered in its own anticipatory self-projection. This projection, a phenonomenology of the one who is dying, is itself temporal, it cannot transcend this horizon of unfathomable finitude. Heidegger writes:

> Dasein never 'finds itself' except as a thrown fact. In the state of mind in which it finds itself, Dasein is assailed by itself as the entity which it still is and already was – that is to say, which it constantly is as having-been.[475]

In this way, an anticipatory resoluteness does not rest in solitude but is a 'moment of vision' of this existence of lived temporality, which can be 'taken back' in light of 'movement' in the phenomenon. However, that which cannot be taken back, that which is beyond all provision, makeshift, is the finitude of existence, of temporality, itself. It is thought, philosophy, expression that is provisional, makeshift, as with that projection 'building plan' in *Kant and the Problem of Metaphysics*. Heidegger writes:

> Ground-laying is rather the projecting of the building plan itself so that it agrees with the direction concerning on what and how the building will be grounded.[476]

The self-projection of finite existence, the being of the self is ec-static transcendence, as the temporal disclosure of this self that uncovers being-in-the-world. Such a thinking is attuned to the phenomenon of temporal be-ing, in all of its uncoverings, disclosures, events. The self temporalizes its own temporality via its resolutions amid its silent predicament. It is in this way that Anaximander speaks to each of us, despite the nuances of translation. Each is finite, each is guilty – for its own free 'insurrection against nothingness', for its own utter recklessness, and the reck that comes around. At the same time, one remembers this 'world' is still thrown, the finitude of existence itself is uncovered as a 'coercion of the phenomena', a 'necessity' to fathom this uncanny finitude of one's existence. It is in this sense that I describe thought, philosophy as a makeshift, as a necessary, but a finite and somehow 'strange' dwelling. In the fragments of Anaximander and other pre-Socratic philosophers, there are extant indications of being-in-the-world, which disclose world as a makeshift ground of finite thinking, the finitude of which is an intimation of its piety to this earth, not a 'power of violence', of a spontaneist projection, but as an 'openness', thrown amid 'phenomena'.

Such a makeshift knowing is indicated by Heidegger in his designation of philosophia (φιλοσοφία) as an *episteme* zetoumene (ἐπιστήμη ζητουμένη): a finite knowing never to become a 'fixed possession' to be merely 'passed on', as

Kant said of his own *arche*. It is instead 'first philosophy', a rough sketch coalescing amid a discordant truth, born anew each day, as a love (φιλειν), a desire (σοφία) for these 'matters themselves'. In this light, philosophy becomes a *first* in the sense of John Cage's score '433', a composition of silence that gives an overwhelming place for chance operations and the expression of that which is there, those who are there, or, as I have written above, philosophy as being-open to the raw 'matters themselves', to the call that 'discourses solely and constantly in the mode of keeping silent'.[477] The reticence over one's self, amid the event, is a decided resistance to the 'detour' of discursivity. In this way, this silence is not that of death, but an openness which allows for self-expression amid the silence, as Schürmann expressed, of thetic fantasms. Being-silent is being-open to the *not-yet* of Dasein in its aletheological birthing, natality. In this way, silence is far from quiet as the self-expression of existence occurs through art, song, poetry and dance – and philosophy intimated as a *makeshift* – as those primitive expressions of the polymorphous truths of our existence.

The Circle of Finitude

Who fights with monsters would like to see that he does not become a monster himself. And when you look long into an abyss, the abyss also looks inside you.[178]

A 'phenomenology of original, ecstatic temporality' is an intimate self-interpretation of one's own temporal, historical existence and its indication and expression via an indigenous conceptuality. Da-sein gathers its meaning amid the horizons of temporality; phenomenology as finite knowing is a 'makeshift', a 'first philosophy', each time taken up anew, never to be a possession which need merely be passed on, only to be applied. Expression is a self-expression of history and the phenomenon of this lived temporality, 'matters themselves'. The phenomena, which are of being, are characterized by tragic difference in the distinction between everydayness and singularity, of being and the over-whelming, in Heidegger's sense. However, the former absorption in the familiar is for Heidegger 'grounded' in an overwhelming, a makeshift ground gained in a provisional 'resolution' of the abyss, amid the eventual recoiling from or destruction by the overwhelming. Finite knowing is 'constituted' via its 'failure' to 'hold itself in the nothing'. It is a flight from an overwhelming, a being over-whelmed, that 'grounds', first indicates, the fragility of finite knowing, thrown into its world, where this sense of its own be-ing is an issue. Philosophy is a ques-tion that reminds us that existence is not simply 'there' but is 'achieved'; senses of an overwhelming are carried back into this 'everyday', a sincere unknowing, one's 'true' perplexity.

In one's resolution amid this abyss of the nothing, one is open for beings *per se*, yet, now as illuminated in these embers of nothingness which remain, which one carries with one. This turning to beings, a falling from the event, is a decay that is yet fertile in that the attempt was made to 'go all the way to the end'. In light of the resolution of the event, Heidegger states that this turning to beings is an overturning (**Umschlag**), *metabole* (μεταβολή), and indicates a *topos* of questions designated as *metontology*. Heidegger states in *The Metaphysical Foundations of Logic*, that the 'turning' from a fundamental ontology to a metontology, or, from the basic analysis of being to the phenomenological apprehension of beings, constitutes the fundamental concept of metaphysics. As we will recall from the *destruktion* of Kant, such a turning points to the dis-tinction of a Metaphysica Generalis and Metaphysica Specialis, or in the context of Aristotle, of *first philosophy* and theology. Metontology, in this way, is not the retreat to a warmed-over empiricism or realism, or a temporal idealism, but an openness towards phenomena in light of an explicit surrender to one's ownmost possibility. A *metontology* arises out of a situation of questioning in conjunction with, and in tandem with, the 'event' of fundamental ontology. After the event, the world takes on a differing light. Heidegger states that it is here, once we have been shattered in the abyss, as we are struggling to return to a world of beings, that a basic comportment of openness can arise amidst the topology of beings. It is upon an ontic fundament, itself setting over an abyss of an overwhelming situation of ecstatic temporality, that an *ethics* is possible, not as with Kant, as a practical reason from which any and all traces of temporality and imagination have been cleansed, but as an *ethos* of finite, free existence, amid the event.

With this light of nothingness carried along back into the world of beings, one already always begins to fall, is falling, and one works to remember this utter finitude of its existence and of its expressions. The indication of a make-shift is intended to point to a thinking which remembers its tenuousness, and seeks, after the example of Reiner Schürmann, a humility of thinking in the awesome face of 'matters themselves'. It is this event of the overwhelming, and one's return from oblivion, which prepares one for an openness for beings, for other people, of faces, hands and voices amid this world. Such a being open, as *gelassenheit,* turns towards the truth of being, as the self-disclosure of temporality and seeks to remain attuned to the phenomenon. Yet, even such attunement does not preclude the uncanny guest of anxiety.

Notes

1   Schürmann, Reiner. *Heidegger: On Being and Acting, from Principles to Anarchy*, p. 29. I have chosen this quotation to frame my primary argument which concerns the question of the first, of the beginning of philosophical investigation as 'first philosophy'.

2   Kisiel, Theodore. *The Genesis of Heidegger's Being and Time*, University of California Press, 1993.

3   For a detailed overview of the primary texts explored in the following pages, see Luchte, J. (2003) 'Makeshift: Phenomenology of Original Temporality', *Philosophy Today*, DePaul University, Vol. 47, No. 3.

4   Most obviously by Kisiel and in the unpublished thesis of Triine Kallas, Krell in *Intimations of Mortality*, Hans Ruin in 'Heidegger in Ruins', Taminiaux in *Heidegger and the Project of Fundamental Ontology*, Sherover in *Heidegger, Kant, and Time*, Schalow in his *The Renewal of the Heidegger-Kant Dialogue*, Yoko Arisaka, in 'Spatiality, Temporality, and the Problem of Foundation in *Being and Time*,' and Beatrice Han's 'Early Heidegger's Appropriation of Kant'.

5   Another issue, as we will see, is the relative neglect by interpreters of the topic of ecstatic temporality and indeed the second division of the published fragment.

6   An important exception to this 'rule' is the posthumous, unfinished, work of Carol White (2005) *Time and Death*, Ashgate, with an Introduction by Hubert Dreyfus. An interesting feature of this work is her attempt to create an interface with some important figures in Analytic philosophy, such as Searle. Perhaps for this reason, she abandons the ecstatic temporality of *Being and Time*, for considerations of time and death in social and cultural contexts.

7   Sherover, in his *Heidegger, Kant and Time*, sets forth an indispensible, but restricted interpretation of *Kant and the Problem of Metaphysics*. There is little to be said of ecstatic temporality, or such, as Sherover is concerned to show that Heidegger's interpretation of Kant must be accurate on the latter's own terms. Yet, as we remain within the 'latter's own terms', we *become* prohibited from entering into the ecstatic *topos* of self-interpretation and expression.

8   *Basic Problems*, p. 23.

9   This is a point that seems to have been missed by commentators such as Blattner, Dreyfus and Carmen as pointed out by Han-Pile in her solid, 'Early Heidegger's Appropriation of Kant.' Yet, she herself seems to miss the *next step* of an explicit endorsement of a radical phenomenology and hermeneutics of existence. This debate seems to be in a perpetual state of postponement.

10 Kiesel claims that such an 'event' was Heidegger's witness to the 'rationality of the trenches' in World War I.

11 As with Aristotle and Brentano, I will express 'being in many ways'. I will avoid the practice of designating Sein, with a capital B, and such, and will allow the nuances

of Sein/sein to express itself, being amid the context or de-context of the situation or event.

[12] Schürmann points out in his *Heidegger* that an *original* temporality, one that founds a world, is to be distinguished from the *originary* temporality of re-leasement, pp. 132ff. I have taken this point in respect to the strategy of reading Heidegger backwards.

[13] This phrasing was suggested by Sherover in his classic (though out of print) *Heidegger, Kant and Time*, as an alternative translation of **Einbildungskraft**, translated in *Kant and the Problem of Metaphysics* by Taft as 'transcendental power of imagination'.

[14] Heidegger explicates this notion in an Appendix to *Metaphysical Foundations of Logic*, pp. 154–59.

[15] Krell, *Intimations of Mortality*.

[16] Heidegger, *Being and Time*, p. 195 (153). This quote was chosen to highlight and to thematize my argument concerning the peculiar character of circularity that is exhibited in Heidegger's phenomenological hermeneutics. An interesting comparison can be made with Schelling's 'On the Possibility of a Form of Philosophy as such,' (1794) in which he writes, 'Here we find ourselves in a magic circle.'

[17] Olafson, F. *Heidegger and the Philosophy of Mind*, p. 78.

[18] This Greek term *topos* which means place, is used as a synonym for the 'Da' of da-sein. Reiner Schürmann, in his *Heidegger on Being and Acting: From Principles to Anarchy* and in his *Broken Hegemonies*, and Frank Schalow in his *The Renewal of the Heidegger-Kant Dialogue: Action, Thought, and Responsibility*, have each insisted on a *topos* (of Being), topology with the former, and topography with the latter. Moreover, Kisiel uses the term *topoi* in the *Introduction* to his *Genesis*.

[19] In Homer's *Iliad*, Alethea is a goddess of childbirth, which is an illuminating sign in the sense of the struggle for unconcealment amid the undertows of concealment.

[20] Safranski, Rudiger (1998) *Martin Heidegger: Between Good and Evil*, Trs. by Ewald Osers, Harvard University Press.

[21] Heidegger, *Basic Problems of Phenomenology*, p. 325. This quotation serves the argumentation of this chapter in that it sets out, as a pre-understanding, the necessity of a differentiation between common or generic time from original temporality.

[22] Schalow, Frank, *The Renewal of the Heidegger-Kant Dialogue*, State University of New York Press, pp. 45–79.

[23] Ibid. p. 50.

[24] Sherover, Charles M. (1988) *Heidegger, Kant, and Time*, Washington D.C.: Center for Advanced Research in Phenomenology & University Press of America.

[25] Schalow, *The Renewal*, p. 53.

[26] *Metaphysical Foundations of Logic*, 127.

[27] Schalow, *The Renewal*, p. 56.

[28] Ibid. 58.

[29] 'On the Unity of Subjectivity', *The Unity of Reason*, p. 18.

[30] For a more schematic juxtaposition of common and original temporality, consult MFL pp. 197–99.

[31] Krell, *Genesis*, p. 442.

[32] *Basic Problems*, 257.

[33] This term is taken from Schürmann's *Broken Hegemonies.*

[34] *Basic Problems*, 257.

[35] Ibid. 258.

[36] This indication of reckoning speaks to the 'reck' in Heidegger's translation of the fragment of Anaximander, as a disclosure of the radical temporality of existence.

[37] *Basic Problems*, 258.

[38] Ibid. 260.

[39] Ibid. 260.

[40] Ibid. 260.

[41] Ibid. 263.

[42] Ibid. 264.

[43] Ibid. 265.

[44] Ibid. 265.

[45] Ibid. 265.

[46] Ibid. 266.

[47] Ibid. 267.

[48] Ibid. 268.

[49] *Being and Time*, 82.

[50] *Basic Problems*, 260.

[51] Ibid. 269.

[52] Ibid. 269.

[53] Ibid. 270.

[54] Ibid. 270.

[55] Ibid. 270.

[56] Ibid. 277.

[57] Ibid. 274.

[58] *Phenomenological Interpretation*, 93.

[59] Ibid. 95.

[60] This is a phrase from Schürmann's 1992 lecture course on Plotinus at the Graduate Faculty, New School University, NYC.

[61] *Phenomenological Interpretation*, 133–36.

[62] *Being and Time,* 61. This quotation serves the argumentation of this chapter as it clearly indicates that Heidegger is seeking a specification of the variation in sources and types of conceptuality. In this case, Heidegger is distinguishing a free-floating, generic conceptuality, from that which arises from an indigenous, temporal context.

[63] This topic of 'ruin' (ruinanz) was explored in an essay by Hans Ruin (Stockholm), 'Heidegger in Ruins,' presented at the University of Essex 'Heidegger Workshop', Spring, 1999.

[64] Taminiaux reports that Heidegger confessed that 'Nietzsche shattered me', *Heidegger and the Project of Fundamental Ontology*, Chapter 6, 'The Presence of Nietzsche in "Sein und Zeit"', pp. 175–91. Kisiel, in his *Genesis*, gives a gripping portrayal of Heidegger's imminent awakening to his own 'death of god'.

[65] Krell, D. *Intimations of Mortality.*

[66] James Joyce. *Finnegan's Wake*, Penguin, New York, 1976. For a recent exploration of the 'relation' between Joyce and Heidegger, see Michael Elfred, 'An Other Beginning – Here Comes Everybody: *The Joycean work of art against the foil of*

Heidegger's thinking,' Conference Paper, Heidegger und die Dichtung, *in Meßkirch 24–28 May 2006.* It is interesting that *Finnegan's Wake* was published at the same time as *Sein und Zeit* and strange that each exhibits its own circularity. It would be in light of this resemblance that Elfred may wish to re-think 'Heidegger' and his alleged 'foil'.

[67] Wittgenstein, Ludwig. *Tractatus Logico-Philosophicus*, 7, p. 151: 'What we cannot speak about we must pass over in silence.' Is the silence of formal indication a language that acts as a showing, as with Wittgenstein's method of phenomeno-logical analysis from the 1930s and beyond? Is it possible to read *Being and Time* as a response to or an elaboration of the *Tractatus*?

[68] History of the Concept of Time, 17.

[69] For a detailed account of the conflict between Heidegger and the logical positiv-ists, cf. Luchte, J. (2007) 'Martin Heidegger and Rudolf Carnap: Radical Phenomenology, Logical Positivism and the Roots of the Continental/Analytic Divide,' *Philosophy Today*, 51, 3.

[70] Ibid. 20.

[71] Ibid. 21.

[72] Ibid. 21.

[73] Ibid. 21.

[74] Ibid. 22.

[75] Ibid. 13.

[76] Husserl, E. *Logical Investigations*, Vol. 2, Routledge, 2001.

[77] *Metaphysical Foundations of Logic*, 160.

[78] History of the Concept of Time, 36.

[79] Ibid. 39.

[80] Ibid. 40.

[81] Ibid. 42.

[82] Ibid. 44.

[83] Ibid. 46.

[84] Ibid. 47.

[85] Ibid. 49.

[86] Ibid. 50.

[87] In this context, 'truth' can have three senses. It can be subsistence of identity betwixt presumed and intuited as perceiving that lives in 'matters themselves', to 'live in the truth'. Also, it can refer to 'evidence', as 'intentio as identifying'. Thirdly, it can point to a being-actual of an entity in a naive existence as it shows itself. In this light, there are two senses of being at play amid this wider sense of truth: on the one hand, being is a truth relation, which intimates Heidegger's retrieval of ontology, and on the other hand, there is the being of a copula, which refers not to an overriding 'intentional' situation, but to subsistence and stasis [Bestand and Stehen] of a being in a truth relation. Corresponding to these dif-fering 'senses' of truth and being, there are two senses of expression.

Indigenous expression is that which 'lives in the truth', 'united', to announce a presence of an act, to be 'animated by it'. Second, there is an expression which merely announces the 'perceived' of an act.

It is in a 'detour' of this latter expression that there is a danger of covering up, or forgetting, original matters.

88 History of the Concept of Time, 60.
89 Ibid. 62.
90 It is significant that in the midst of his location of the *topos* of the 'matters themselves', Heidegger makes a direct reference to Plato's *Sophist*, the text with which he begins *Being and Time*.
91 History of the Concept of Time, 73.
92 Ibid. 74.
93 Ibid. 77.
94 Ibid. 77.
95 Ibid. 78.
96 Ibid. 78.
97 Ibid. 84.
98 Ibid. 84.
99 Ibid. 85.
100 Ibid. 85.
101 Ibid. 86.
102 Ibid. 85.
103 Ibid. 87.
104 Ibid. 87.
105 Heidegger warns that we must therefore resist a temptation to 'place' pictures amid an artificial context, as for instance, with Hegel into a 'phenomenology of spirit', or to obliterate this 'what' of phenomena, as Husserl has done, into a 'consciousness'.
106 History of the Concept of Time, 95.
107 Ibid. 97.
108 Ibid. 97.
109 Ibid. 98.
110 Ibid. 99.
111 Ibid. 100.
112 Ibid. 100.
113 Ibid. 101.
114 Ibid. 101.
115 Ibid. 106.
116 Ibid. 107.
117 Ibid. 110.
118 In this regard, Gadamer's criticism in *Truth and Method* of Dilthey's commitment to scientific method can cast some light upon an interpretation of the relation of Heidegger and Dilthey's 'life-philosophy'. In the end, Heidegger does not use the word 'life', but Dasein as at once ecstatic and neutral, in-between. Gadamer wishes, with Heidegger, to articulate a *topos* with its own hermeneutical orientation and language.
119 History of the Concept of Time, 127.
120 Ibid. 129.
121 Ibid. 131.
122 Ibid. 136.
123 Ibid. 137.
124 Ibid. 137.

[125] Ibid. 141.

[126] Ibid. 143.

[127] Ibid. 144.

[128] Ibid. 146.

[129] Ibid. 147.

[130] Ibid. 148.

[131] George Bataille, *Inner Experience*, p. 95. This quotation serves the argumentation of this chapter as it lays out the target of Heidegger's thematization of being-in-the-world (In-der-Welt-sein), which is, in *Being and Time*, the isolated Cartesian and Leibnizian mental substances.

[132] *Basic Problems*, 277.

[133] History of the Concept of Time, 165.

[134] History of the Concept of Time, 233.

[135] Lefebvre in his *The Production of Space* points to Heidegger as being 'first among philosophers' to explore these 'vicissitudes' of the relations between space, place (*topos*) and language, p. 242. However, he criticizes Heidegger's priority of time over space. Yet, in this way, it is uncertain if the richness of Heidegger's indication of the existential spatiality of dasein was ever noticed by Lefebvre.

[136] This term comes from George Bataille's *Inner Experience*.

[137] Sallis, J. *Echoes – After Heidegger*. Indiana University Press, 1990.

[138] Piero Sraffa, *The Production of Commodities by Means of Commodities*, Cambridge University Press, 1979. Sraffa, a 'post-Keynesian' economist, was friendly with Keynes and Antonio Gramsci and was a fierce opponent of Mussolini.

[139] History of the Concept of Time, 153.

[140] Author's translation from Friedrich Nietzsche (1967) Jenseits Von Gut Und Böse: Vorspiel einer Philosophie der Zunkunft, Werke in Zwei Banden, II, Munich: Carl Hanser Verlag, p. 61.

[141] The prospective-retrocursive structure is suggested by picture drawn of the projections of ecstatic temporality during Heidegger's 1928 lecture course, *The Metaphysical Foundations of Logic*, in which he describes the 'backward and forward' movements of the 'hermeneutical circle', p. 206.

[142] *Basic Problems*, 275.

[143] Ibid. 276.

[144] Ibid. 276.

[145] Ibid. 277.

[146] Ibid. 278.

[147] Ibid. 279.

[148] Taylor Carmen taunts us with the possibility of a 'virtue of inauthenticity', in his 'Must we be Inauthentic?' in *Heidegger, Authenticity and Modernity*, Cambridge, 2000.

[149] *Basic Problems*, 280.

[150] Ibid. 280.

[151] Ibid. 280.

[152] Heidegger points out that in the 'Myth of the Cave' from the *Republic*, Plato refers to this light as the Good (ἀγαθός), the Sun (ελιος), this god who tells Demeter the location of her daughter Persephone in Tartarus. As he is all-seeing (πανοπσις), ελιος is also the true (ἀληθής).

¹⁵³ *Basic Problems*, 285.

¹⁵⁴ Ibid. 286–87.

¹⁵⁵ Ibid. 289.

¹⁵⁶ Ibid. 287.

¹⁵⁷ Ibid. 291.

¹⁵⁸ Ibid. 294.

¹⁵⁹ Ibid. 297.

¹⁶⁰ Ibid. 300.

¹⁶¹ Ibid. 303.

¹⁶² Ibid. 304.

¹⁶³ *Praesens* is not, in this usage, the same as 'praesens' as the horizon (**woraufhin**) of the ecstasis of the present (the use of the Latin 'praesens', Heidegger informs us, distinguishes the current discussion of temporality from that of the earlier discussion which was still 'indifferent, unarticulate . . .' of the present, of a 'now').

¹⁶⁴ Derrida points out in his *Ousia and Gramme*, such a metaphysics of presence is only possible on such a logical interpretation of time, of discrete identities of the plethora of 'nows'. However, it is not evident that Derrida is willing to consider Heidegger's tenative intimations with respect to *Praesens*. In that he had gone out of his way to speak of the Nothing (**das Nichts**), a closer look at his lectures is warranted, which Derrida did on other occasions, as in the case of sexual difference.

¹⁶⁵ *Basic Problems*, 306.

¹⁶⁶ Ibid. 307.

¹⁶⁷ Ibid. 307.

¹⁶⁸ Ibid. 307.

¹⁶⁹ Ibid. 308.

¹⁷⁰ Ibid. 308.

¹⁷¹ This is an indication expressed by Reiner Schürmann in his lectures on Plotinus at the Graduate Faculty, New School for Social Research, Autumn, 1992.

¹⁷² Taminiaux, J. *Heidegger and the Project of Fundamental Ontology*, p. 68. Taminiaux writes: 'Its deconstruction is not exposed to an outside, to a "difference," to a semantic dispersion.' He refers to Derrida's *Speech and Phenomena*, suggesting that Heidegger's 'destruktion' which, in its preference for violence over play, inevitably remains tied to a 'metaphysics of subjectivity'. Derrida addresses the question of *destruktion* and deconstruction in 'Differance', 'Ousia and Gramme: Note from a Note from *Being and Time*', and 'The Ends of Man'. In 'Ousia and Gramme', Derrida articulates the question of whether or not the distinctions written into Heidegger's works of fundamental ontology, between 'authenticity' and 'inauthenticity', 'primordial' and 'common' are merely repetitions of the violence exacted by a 'metaphysics of presence'. It is not clear however whether Heidegger's use of the term praesens can be so readily assimilated to the 'history of ontology'.

¹⁷³ Weatherston, Martin (2002) *Heidegger's Interpretation of Kant: Categories, Imagination, and Temporality*, New York: Macmillan.

¹⁷⁴ Michel Foucault, *Discipline and Punish*, p. 197. This quotation serves the argumentation of this chapter as it lays out an appropriate model of the systematic structure of rational discipline which is the target of Heidegger's *destruktion*.

175 Compare this to the non-relational character of being-towards-death.

176 *Basic Problems*, 47. This metaphoricity of the 'monstrous site' recurs in Schürmann's *Broken Hegemonies*.

177 This is an indication from Reiner Schürmann's *Heidegger: On Being and Acting, From Principles to Anarchy* in which he criticizes Derrida's deployment in *Writing and Difference* of the table of concepts to characterize Heidegger's principles of the epochal dispensations of Being.

178 *Basic Problems*, 47.

179 Ibid. 314.

180 Ibid. 315.

181 Ibid. 316. The citation is from the Academy Edition, vol. 17 (vol. 4 of div. 3), No. 4017, p. 387. [Immanuel Kant, *Gesammelte Schriften* (Berlin and New York: W. de Gruyter, 1902).]

182 Ibid. 316.

183 Once again, Heidegger does not set out the horizons of the ecstases of the future and having been, in the sense of a futurens or a passens, which is tentatively indicated in Chapter 9.

184 *Basic Problems*, 319.

185 Ibid. 319.

186 Bataille, *Theory of Religion*, 50.

187 In this light, one can minimize the importance of Adorno's rather sloppy rant against what he documented as a penetration of existentialist terminology into popular culture at the close of the 1960s. His *Jargon of Authenticity*, as he points out, is more of an attack upon non-philosophers, than upon Heidegger per se. For Adorno, the fact that such a seizing upon of existentialist language occurred at all, indicates a crossing of a line and a 'dangerous' potential to not only philosophy, but for culture as such. For him, Heidegger's work is suspect as it transgresses the 'limits' of the forbidden, or in other words, the limits of the 'real' – that it erupted into its *world*. But, with Nietzsche, why should we not seek to be and be dangerous? Consider the case of Socrates . . .

188 Bataille emphasizes the collective character of dasein in an extended reference to Heidegger, beginning on page 24 of *Inner Experience*. For an account of the explicit and implicit relations between Heidegger and Bataille, see Rebecca Comay, 'Gifts without Presents: Economies of "Experience" in Bataille and Heidegger', in *Yale French Studies: On Bataille*, No. 78, pp. 66–89.

189 *History of the Concept of Time*, 272.

190 Ibid. 272.

191 *Metaphysical Foundations of Logic*, 106.

192 Ibid. 71. This quotation serves the argumentation of this chapter as it clearly shows that Heidegger's interpretation has no intention of seeking a consensus, but will be an originary appropriation of the philosophy of Kant, articulated from the context of a resolute interpretation.

193 Sherover, Charles, *Heidegger, Kant, and Time*, p. 147.

194 *Being and Time*, 45.

195 This expression is inspired by Goethe's Botanical essay, 'On the Metamorphosis of Plants', which can be described as a very early descriptive phenomenology or morphology of metamorphic growth.

196 Cf. Taminiaux's emphasis on conscience as a 'hearing of the self' – as a voice which gathers. Also Giorgio Agamden's lecture course, *Language and Death: The Place of Negativity*, on the voice and temporality.

197 John Sallis, in his *Echoes – After Heidegger*, calls for a re-inscription of imagination into phenomenology as an indication of a fictive paradox of 'returning to the things themselves', p. 97ff.

198 *Being and Time*, 45.

199 Ibid. 45.

200 'Unity of Subjectivity' in *The Unity of Reason*.

201 *Kant and the Problem of Metaphysics*, Section 35, p. 137 and *Kant und das Problem of Metaphysik*, pp. 195ff.

202 Ernst Cassirer discusses Heidegger's Kantbook in 'The Problem of Metaphysics', *Kant: Disputed Questions*, p. 149.

203 'The Problem of Metaphysics', a review of *Kant and the Problem of Metaphysics*, in *Kant: Disputed Questions*, p. 154.

204 The Unity of Subjectivity', *The Unity of Reason*, p. 17ff.

205 Ibid. 17.

206 Ibid. 13.

207 Ibid. 15.

208 Ibid. 15.

209 Ibid. 15–16.

210 Ibid. 21.

211 Sherover, Charles, *Heidegger, Kant and Time*.

212 *Kant and the Problem of Metaphysics*, 17.

213 Ibid. 108.

214 Ibid. 19.

215 Ibid. 21.

216 Ibid. 24.

217 Ibid. 25.

218 Ibid. 27.

219 Ibid. 28.

220 Ibid. 29.

221 Ibid. 30.

222 *Metaphysical Foundations of Logic*, 210–11. This quotation serves the argumentation of this chapter as it intimates the dominant claims that characterize Heidegger's treatment of Kant. From this perspective, we are forced to re-evaluate the meaning of interpretation and the purpose of critical philosophy.

223 *Kant and the Problem of Metaphysics*, 30.

224 Ibid. 34.

225 *Critique of Pure Reason*, A34, B50; KPM 34.

226 *Kant and the Problem of Metaphysics*, 35.

227 *Phenomenological Interpretation*, 103.

228 *Kant and the Problem of Metaphysics*, 35.

229 Ibid. 37.

230 Ibid. 38.

231 Ibid. 38.

232 Ibid. 38.

233 Ibid. 38.
234 Ibid. 41.
235 Ibid. 43.
236 Ibid. 43.
237 Ibid. 43.
238 Ibid. 44.
239 Ibid. 44.
240 Kant here follows Aristotle's definition of imagination as a mediating in *De Anima.*
241 'The Unity of Subjectivity', p. 48ff.
242 *Kant and the Problem of Metaphysics*, 46.
243 Ibid. 46.
244 Ibid. 46.
245 Ibid. 48.
246 'Heidegger in Ruins', presented by Hans Ruin of Stockholm at the 'Heidegger Workshop' at Essex (1999), organized by Simon Critchley, to exhibit the background of ruin to Heidegger's philosophy.
247 *Kant and the Problem of Metaphysics*, 48.
248 Ibid. 49.
249 Ibid. 49.
250 Ibid. 50.
251 Ibid. 50.
252 Ibid. 51.
253 Ibid. 52.
254 Ibid. 52.
255 Ibid. 53.
256 Ibid. 53.
257 Ibid. 54.
258 Ibid. 56.
259 Ibid. 56.
260 Ibid. 56.
261 Ibid. 57.
262 Ibid. 57.
263 Ibid. 58. Cf. *Basic Problems* on imagination and fantasy.
264 Ibid. 59.
265 Ibid. 59.
266 Ibid. 62.
267 Ibid. 67.
268 Ibid. 63.
269 Ibid. 64.
270 Ibid. 64.
271 Ibid. 66.
272 Ibid. 67.
273 Ibid. 67.
274 Ibid. 67.
275 Ibid. 68.
276 Ibid. 69.

277 Ibid. 70.

278 Ibid. 71.

279 Ibid. 73.

280 Ibid. 73.

281 Ibid. 75.

282 Heidegger refers to this Nietzschean myth in the opening pages of his 1925 lecture course, *History of the Concept of Time.*

283 *Kant and the Problem of Metaphysics*, 77.

284 Ibid. 78.

285 Ibid. 78.

286 Heidegger quotes a 1797 letter of Kant: 'In general, the Schematism is one of the most difficult points. Even Herr Beck cannot find his way therein. – I hold this chapter to be one of the most important' (KPM 80).

287 *Kant and the Problem of Metaphysics*, 80.

288 Ibid. 81.

289 Ibid. 84.

290 Ibid. 84.

291 *Critique of Pure Reason*, A155, B194.

292 *Kant and the Problem of Metaphysics*, 87.

293 Ibid. 87.

294 Ibid. 88.

295 Descartes, writing to Picot, quoted by Heidegger at the head of his essay, 'The Way Back into the Ground of Metaphysics', *Existentialism*, edited by Kaufmann, p. 207.

296 *Basic Problems*, 86.

297 Ibid. 90.

298 *Phenomenological Interpretation*, 41.

299 *Kant and the Problem of Metaphysics*, 90.

300 Ibid. 91. Imagination is also mentioned in this regard in the B Deduction of the First *Critique* as a term of disparagement, an indication which would fit well with Heidegger's argument.

301 Ibid. 91.

302 Ibid. 91.

303 Ibid. 91.

304 Ibid. 92.

305 Ibid. 93.

306 Ibid. 94.

307 Ibid. 94.

308 Ibid. 95.

309 Sallis, John. *Spacings – of Reason and Imagination*, University of Chicago Press, 1987.

310 *Kant and the Problem of Metaphysics*, 96.

311 Heidegger referred to philosophy as the 'atrophied root of the science' in a 1929 lecture, 'The Essence of Truth', an essay which is given a twenty-year commemoration lecture in 'On the Way Back into the Ground of Metaphysics' in Kaufmann's *Existentialism*. In this lecture, Heidegger insists we 'stick' to the schema image of the 'common root' since it indicates a stepping-back.

[312] *Kant and the Problem of Metaphysics*, 98.
[313] Ibid. 98.
[314] Ibid. 101.
[315] Ibid. 99.
[316] Ibid. 100.
[317] Ibid. 103.
[318] Ibid. 103.
[319] Ibid. 104.
[320] Ibid. 105.
[321] Ibid. 105.
[322] Ibid. 105.
[323] Ibid. 106.
[324] Ibid. 106.
[325] Ibid. 109.
[326] Ibid. 110–11.
[327] Ibid. 109.
[328] Ibid. 110.
[329] Ibid. 112.
[330] Ibid. 112.
[331] Zizek, Slavoj. 'The Deadlock of Transcendental Imagination, or, Martin Heidegger as a Reader of Kant,' *The Ticklish Subject*, pp. 9–69, specifically pp. 42ff.
[332] Sallis, John. (1990) *Echoes – After Heidegger*. Bloomington: Indiana University Press.
[333] *Kant and the Problem of Metaphysics*, 112.
[334] Ibid. 117.
[335] Ibid. 119.
[336] Ibid. 198. This quotation serves the argumentation of this chapter as it underlines the necessity of disclosing the *a priori* temporal horizon that serves as the context of emergence for any consideration of ontological status or character.
[337] Makkreel, *Imagination and Interpretation in Kant*, 21.
[338] *Kant and the Problem of Metaphysics*, 121.
[339] Ibid. 123.
[340] Ibid. 123.
[341] Ibid. 123.
[342] Ibid. 124.
[343] Ibid. 126.
[344] Ibid. 126.
[345] Ibid. 126.
[346] Ibid. 127.
[347] Ibid. 127.
[348] Ibid. 129.
[349] Ibid. 130.
[350] Ibid. 130.
[351] Ibid. 130.
[352] Ibid. 130.
[353] Ibid. 131.
[354] Ibid. 132.

[355] Ibid. 132.

[356] *Aristotle Rhetorik: Summer 1924*, as discussed by Kisiel in *The Genesis of Heidegger's Being and Time*, p. 298.

[357] *Kant and the Problem of Metaphysics*, 132.

[358] Ibid. 133.

[359] Ibid. 135.

[360] Ibid. 141.

[361] Ibid. 142.

[362] Ibid. 170.

[363] Ibid. 170.

[364] Heidegger, M. (1972) 'On Time and Being', New York: Harper Torchbooks.

[365] Kisiel, *Genesis*, 498.

[366] *Basic Problems*, p. 101. This quotation serves the argumentation of this chapter as it allows us to thematize the peculiar mine-ness of a resolute self-interpretation of Dasein. In this way, any consideration of other philosophers is already undertaken in the context of a projection of authentic self-interpretation.

[367] Cf. Luchte, J. (2006) 'Mathesis and Analysis: Finitude and the Infinite in the *Monadology* of Leibniz', London: *Heythrop Journal*.

[368] Bataille, George, *Inner Experience* (1954), p. 129. This quotation serves the argumentation of this chapter as it thematizes the 'inauthenticity' of interpretations of existence which rely upon the 'dead language' of logic, which, though is our creation, serves to conceal the truth of our being.

[369] *Metaphysical Foundations of Logic*, 8.

[370] Ibid. 5.

[371] Ibid. 5.

[372] Ibid. 5.

[373] Ibid. 9.

[374] Ibid. 10.

[375] Ibid. 11.

[376] Ibid. 11.

[377] Ibid. 12.

[378] Ibid. 17.

[379] Kant, CPR B140–41. Quoted in MFL 30.

[380] *Metaphysical Foundations of Logic*, 32.

[381] The term 'subject' has three senses, which parallels the stems-root schema in Heidegger: an ontic substance, or monad, and a logical 'subject' of the statement as 'stems' and a pre-eminent subject, or 'I', the authentic self as 'root'.

[382] *Metaphysical Foundations of Logic*, 34.

[383] Ibid. 49.

[384] Ibid. 55.

[385] Ibid. 62.

[386] Ibid. 82.

[387] Ibid. 86. Quotation of a letter to de Volder, 1705.

[388] Ibid. 88.

[389] Ibid. 88.

[390] Kirk, G. S. et al., p. 198, Fragment 220.

[391] *Metaphysical Foundations of Logic*, 90.

392 Ibid. 102.

393 Ibid. 102.

394 *Basic Problems*, 300.

395 Ibid. 301.

396 Heraclitus, Fr. 67, Kirk, p. 190.

397 *Metaphysical Foundations of Logic*, 115.

398 Ibid. 123.

399 Ibid. 123.

400 Ibid. 125.

401 Odo Marquard discusses 'plurivocity' in his article, 'In Praise of Polytheism', where he praises polytheism for its respect of these myriad voices of truth.

402 *Metaphysical Foundations of Logic*, 134.

403 Ibid. 142.

404 Ibid. 142.

405 Ibid. 144.

406 Ibid. 147.

407 Ibid. 163.

408 Ibid. 165.

409 Ibid. 165.

410 Ibid. 171.

411 Ibid. 172.

412 Ibid. 175. Cf. Schalow on world in his text, *The Renewal of the Heidegger-Kant Dialogue*, in the section, entitled, 'The Subterranean Concern of the Dialectic', pp. 91–107. A *transcendental idea* of world, as a transcendental 'ideal', cannot serve as the seat of absolute causation, and is thus outside of the domain of a causative, ontical series.

413 Ibid. 182.

414 Ibid. 182.

415 Ibid. 183.

416 Ibid. 184.

417 Ibid. 185.

418 Ibid. 185.

419 Ibid. 186.

420 Ibid. 187.

421 Ibid. 188.

422 Bakünin, Michael. 'The Illusion of Universal Suffrage', *The Anarchist Reader*, pp. 108–09.

423 *Metaphysical Foundations of Logic*, 193.

424 Ibid. 194.

425 Ibid. 200.

426 Ibid. 201.

427 Ibid. 206.

428 Ibid. 205.

429 Ibid. 205.

430 Ibid. 206.

431 Ibid. 207.

432 Ibid. 208.

[133] Ibid. 208.

[134] Ibid. 208.

[135] Ibid. 209. For an etymology of this term, cf. MFL 209.

[136] Ibid. 210.

[137] Ibid. 214.

[138] Bataille, George, *Theory of Religion*, 11.

[139] *Being and Time*, 28.

[140] Macquarrie, in his translation of the larger fragment of *Being and Time*, misleadingly translates **Befindlichkeit** as 'state of mind'. In this light, I have deferred to Kisiel's translations in *History of the Concept of Time*, p. 255, as disposition and self-finding, which he also discusses in his compendium, *Genesis of Heidegger's Being and Time*.

[141] *Being and Time*, 182.

[142] Caputo, J. *Radical Hermeneutics: Repetition, Deconstruction and the Hermeneutic Project*, Indiana University Press, 1987.

[143] *Being and Time*, 255.

[144] Bataille, Georges, *Inner Experience*, p. 115.

[145] *Basic Problems*, 269.

[146] *Metaphysical Foundations of Logic*, 10. This quotation serves the argumentation of this chapter as it underscores the radical temporalization of thought disclosed in Heidegger's early philosophy, and also serves as the basis therefore for an understanding of the meaning of anticipatory resoluteness.

[147] Schürmann, R. (2003) *Broken Hegemonies*, IUP.

[148] In Heidegger's 'The Anaximander Fragment', p. 57, he displaces the 'normal' translation via a re-translation of each 'Greek' figure, giving a differing and unfamiliar sense to the 'Greek' text. For instance, he sets out the following translation of the 'presumably genuine text' which excludes the entire traditional first phrase:

. . . κατα το χρεων • διδοσαι γαρ αυτα δικην και τισιν ἀλλήλοις της ἀδικίας

. . . along the lines of usage; for they let order and thereby also reck belong to one another (in the surmounting) of disorder.

The ordinary translation, which is taken from Nietzsche is:

. . . according to necessity; for they must pay penalty and be judged for their injustice . . .

[149] Wittgenstein, Ludwig, *Tractatus Logico-Philosophicus*, p. 3.

[150] Nietzsche, Friedrich, *Beyond Good and Evil*, § 1–4.

[151] *Being and Time*, 282.

[152] Ibid. 283.

[153] Ibid. 284.

[154] Ibid. 294.

[155] Ibid. 298.

[156] Ibid. 303.

[157] Ibid. 308.

[158] Ibid. In this context, Heidegger refers to Nietzsche's dictum of not being 'too old for one's victories'.

[159] Ibid. 312.

[160] Ibid. 315.

461 Ibid. 316.
462 Ibid. 318.
463 Ibid. 320.
464 Ibid. 320.
465 Ibid. 320.
466 Ibid. 323.
467 Ibid. 345.
468 Ibid. 52.
469 Ibid. 355.
470 Ibid. 359.
471 Ibid. 363.
472 Ibid. 368.
473 Ibid. 371.
474 Levinas, *Otherwise than Being*, p. 17.
475 *Being and Time*, 376.
476 *Kant and the Problem of Metaphysics*, 2.
477 *Being and Time*, 318.
478 Author's translation of Friedrich Nietzsche (1967) Jenseits Von Gut Und Böse: Vorspiel einer Philosophie der Zunkunft, Werke in Zwei Banden, II, Munich: Carl Hanser Verlag, p. 70.

Bibliography and Further Reading

I. Works by Martin Heidegger

(1999) 'Letter on Humanism', *Basic Writings*, London: Routledge.
(1998) *Kant und das Problem der Metaphysik*, Frankfurt Am Main: Vittorio Klostermann.
(1997) *Kant and the Problem of Metaphysics*, Trs. by Richard Taft. Bloomington: Indiana University Press.
(1997) *Phenomenological Interpretation of Kant's Critique of Pure Reason*, Trs. by Parvis Emad & Kenneth Maly, Bloomington: Indiana University Press.
(1993) *Sein und Zeit*. Tubingen: Max Niemeyer Verlag.
(1992) *The Concept of Time*, Oxford: Basil Blackwell.
(1992) *History of the Concept of Time*, Trs. by Theodore Kisiel, Bloomington: Indiana University Press.
(1992) *Metaphysical Foundations of Logic*, Trs. by Michael Heim, Bloomington: Indiana University Press.
(1988) *Basic Problems of Phenomenology*, Trs. by Albert Hofstadter, Bloomington: Indiana University Press.
(1984) 'The Anaximander Fragment', *Early Greek Thinking*, Trs. by David Farrell Krell and Frank A. Capuzzi, San Francisco: Harper & Row.
(1977) *The Question Concerning Technology*, Trs. by W. Lovitt, New York: Harper Torchbooks.
(1975) *Poetry, Language, Thought*, Trs. by Albert Hofstadter, New York: Harper and Row.
(1973) 'Kant's Thesis about Being', Trs. by Ted E. Klein and William E. Pohl. *Southwestern Journal of Philosophy*, 4 : 7–33.
(1972) 'On Time and Being', *Time and Being*, New York: Harper Torchbooks.
(1967) *What is a Thing?* Trs. by W. B. Barton and Vera Deutsch. Chicago: Regnery.
(1962) *Being and Time*, Trs. by John Macquarrie and Edward Robinson, New York: Harper & Row Publishers.

II. Other sources and references

Adorno, Theodor (1973) *The Jargon of Authenticity*, Trs. by Knut Tanowski & Frederic Will, London: Routledge & Kegan Paul.
Agamden, Giorgio (1991) *Language and Death: The Place of Negativity*, Trs. by Karen E. Pinkus, University of Minnesota Press.

Allison, Henry E. (1983) *Kant's Transcendental Idealism*, New Haven and London: Yale University Press.

Aristotle (1986) *De Anima (On the Soul)*, New York, Penguin.

Bakünin, Michael(1986) 'The Illusion of Universal Suffrage', *Anarchist Reader*, ed. George Woodcock, Fontana Press, pp. 108–09.

Bataille, George (1992) *Theory of Religion*, Trs. by Robert Hurley, New York: Zone Books.

— (1988) *Inner Experience*, Trs. by Leslie Anne Boldt. Albany: State University of New York Press.

Beiser, Frederick C. (1987) *The Fate of Reason*, Cambridge: Harvard University Press.

Blattner, W. (1999) *Heidegger's Temporal Idealism*, Cambridge: Cambridge University Press.

Brentano, Franz (1988) *Philosophical Investigations on Space, Time, and the Continuum*, Trs. by Barry Smith, London: Croom Helm.

Caputo, J. (1987) *Radical Hermeneutics: Repetition, Deconstruction and the Hermeneutic Project*, Bloomington: Indiana University Press.

Cassirer, Ernst (1967) 'Remarks on Martin Heidegger's Interpretation of Kant', *Kant: Disputed Questions*, ed. Molke S. Gram, Chicago: Quadrangle Books.

Comay, Rebecca (1990) 'Gifts without Presents: Economies of "Experience" in Bataille and Heidegger', *Yale French Studies: On Bataille*, No. 78, Yale University Press, pp. 66–89.

Dallmayr, Fred (1993) *The Other Heidegger*, London: Cornell University Press.

Dastur, Francoise (1996) 'The Ekstatico-horizonal Constitution of Temporality', *Critical Heidegger*, ed. Christopher Macann, London: Routledge.

Derrida, Jacques (1986) 'Différance', *Margins of Philosophy*, Trs. by Alan Bass. Chicago: University of Chicago Press.

— (1986) 'The Ends of Man', *Margins of Philosophy*, Trs. by Alan Bass. University of Chicago Press.

— (1984) 'Ousia and Grammé: A Note on a Note from *Being and Time*', *Margins of Philosophy*, Trs. by Alan Bass. Chicago: University of Chicago Press.

— (1973) *Speech and Phenomena*, Trs. by David B. Allison, Evanston: North Western University Press.

Dreyfus, Hubert (1997) *Being in the World: Commentary on Heidegger's Being and Time, Division I*, MIT.

— (1979) *Discipline and Punish*, New York: Vintage Books.

Fynsk, Christopher(1993) *Heidegger: Thought and Historicity*, Cornell University Press.

Gadamer, Hans Georg (1989) *Truth and Method*, London: Continuum Publishing.

Goethe, J. W.(1946) 'An Attempt to Interpret the Metamorphosis of Plants', *Chronica Botanica*, Vol. 10, No. 2, pp. 89–115.

Han-Pile, B. (2003) 'Early Heidegger's Appropriation of Kant', *A Companion to Heidegger*, eds. H. Dreyfus and M. Wrathall, London: Basil Blackwell.

Henrich, Dieter (1994) 'The Unity of Subjectivity', *The Unity of Reason*, Cambridge: Harvard University Press.

— (1971) *The Phenomenology of Internal Time-Consciousness*, ed. by Martin Heidegger, Trs. by James S. Churchill, Bloomington: Indiana University Press.

— (1970) *Logical Investigations*, Trs. by J. N. Findley, New York: Routledge & Kegan Paul.

— (1962) *Ideas*, Trs. by W. R. Boyce Gibson, New York: Collier.

Joyce, James (1976) *Finnegans Wake*, New York: Penguin.

Kant, Immanuel (1996) *Critique of Practical Reason*, Trs. by T. K. Abbott. New York: Prometheus Books.

— (1974) *Anthropology from a Pragmatic Point of View*, Trs. by Mary J. Gregor. The Hague, Matinu Nijhoff.

—(1965) *Critique of Pure Reason*, Trs. by Norman Kemp Smith, New York: St. Martin's Press.

— (1960) *Religion within the Limits of Reason Alone*, Trs. by Theodore M. Greene & Hoyt H. Hudson, New York: Harper Torchbooks.

— (1951) *Critique of Judgement*, Trs. by J. H. Bernard, New York: Hafner Press.

Kaufmann, Walter (1962) *Existentialism*, Cleveland: Meridian.

Kirk, G. S., J. E. Raven, M. Schofield (1988) *Presocratic Philosophers*, Cambridge: Cambridge University Press.

Kisiel, Theodore (1993) *The Genesis of Heidegger's Being and Time*, Berkeley: University of California Press.

Kisiel, T. & John Van Buren (1994) *Reading Heidegger from the Start: Essays in His Early Thought*, Albany: State University of New York Press.

Krell, David Farrell (1986) *Intimations of Mortality*, Pittsburg: Pennsylvania State University Press.

Lawlor, Leonard (1992) *Imagination and Chance*, Albany: State University of New York.

Lefebvre, Henri (1991) *The Production of Space*, Trs. by Donald Nicholson-Smith, Blackwell.

Leibniz, G. W. (1990) *Monadology*, Trs. by G. Montgomery, LaSalle: Open Court.

Levinas, Emmanuel (1998) *Otherwise Than Being*, Trs. by Alphonso Lingus, Pittsburgh: Duquesne University Press.

Macann, Christopher (1996) 'Heidegger's Kant Interpretation', *Critical Heidegger*, London: Routledge, pp. 97–120.

Marquard, Odo (1989) 'In Praise of Polytheism (On Monomythical and Polymythical Thinking)', *Farewell to Matters of Principle*, Oxford: Oxford University Press.

Makkreel, Rudolf A. (1990) *Imagination and Interpretation in Kant*, Chicago: University of Chicago Press.

Marx, Werner (1971) *Heidegger and the Tradition*, Evanston: Northwestern University Press.

May, Reinhard (1996) *Heidegger's Hidden Sources*, Trs. by Graham Parkes, Routledge, London.

Nietzsche, Fredrich (1995) *Untimely Meditations*, Trs. by R. J. Hollingdale, Cambridge: Cambridge University Press.

— (1988) *Beyond Good and Evil: Prelude to a Philosophy of the Future*, Trs. by R. J. Hollingdale, London: Penguin.

— (1974) *Gay Science*, London: Vintage Books.

Olafson, Frederick A. (1987) *Heidegger and the Philosophy of Mind*, New Haven and London: Yale University Press.

Patočka, Jan (2002) *Plato and Europe*, Trs. by Petr Lom, Stanford: Stanford University Press.

Roth, Michael (1996) *The Poetics of Resistance: Heidegger's Line*, Evanston Northwestern University Press.

Safranski, Rudiger (1998) *Martin Heidegger: Between Good and Evil*, Trs. by Ewald Osers, Cambridge: Harvard University Press.

Sallis, John (1990) *Echoes – After Heidegger*, Bloomington: Indiana University Press.

— (1987) *Spacings – of Reason and Imagination*, Chicago: University of Chicago Press.

Schalow, Frank (1992) *The Renewal of the Heidegger-Kant Dialogue*, Albany: State University of New York Press.

(1986) *Imagination and Existence: Heidegger's Retrieval of the Kantian Ethic*, University Press of America, Lanhm, MD.

Schürmann, Reiner (2003) *Broken Hegemonies*, Trs. by Reginald Lilly, Bloomington: Indiana University Press.

— (1987) *Heidegger: On Being and Acting, From Principles to Anarchy*, Bloomington: Indiana University Press.

— (1978) *Meister Eckhart: Mystic and Philosopher*, Bloomington: Indiana University Press.

Sherover, Charles M. *Heidegger, Kant, and Time* (1988) Washington D.C.: Center for Advanced Research in Phenomenology & University Press of America.

— (1988) 'The Hermeneutic Structure of Resoluteness: A Preliminary Exploration', *Hermeneutic Phenomenology*, Washington, D.C.: Center for Advanced Research in Phenomenology and University Press of America.

Sikka, Sonya (1997) *Forms of Transcendence: Heidegger and Medieval Mystical Theology*, Albany: State University of New York Press.

Sraffa, Piero (1979) *Production of Commodities by Means of Commodities*, Cambridge: Cambridge University Press.

Taminiaux, Jacques (1991) *Heidegger and the Project of Fundamental Ontology*, Albany: State University of New York Press.

Volpi, Franco (1996) 'Dasein as Praxis: The Heideggerian Assimilation and the Radicalization of the Practical Philosophy of Aristotle', *Critical Heidegger*, London: Routledge pp. 27–66.

Vycinas, Vincent (1961) *Earth and Gods*. The Hague: Martinus Nijhoff.

Warnock, Mary (1976) *Imagination*, London: Faber and Faber.

Weatherston, Martin (2002) *Heidegger's Interpretation of Kant: Categories, Imagination, and Temporality*, New York: Macmillan.

White, Carol(2005) *Time and Death,*) London: Ashgate.

Zimmerman, Michael (1990) *Heidegger's Confrontation with Modernity*, Bloomington: Indiana University Press.

Zizek, Slavoj (1999) *The Ticklish Subject*, London: Verso.

Index

16096955R00113

Printed in Great Britain
by Amazon